Alan M.

TURING

≋

"an even-tempered, lovable character with an impish sense of humour and a
modesty proof against all achievement."

"He thought so little of physical discomfort that he did not seem to apprehend
in the least degree why we felt concerned about him, and refused all help."

"in a short life he accomplished much, and to the roll of great names in the
history of his particular studies added his own."

∼

So is described one of the greatest figures of the 20th century, yet someone who
was barely known beyond mathematical corridors till the revelations in the
1970s. It was then that Alan Turing's critical contributions to the breaking
of the German Enigma code and the development of computer science, along
with the circumstances of his suicide at the height of
his powers, became widely known.

From the rather odd, precocious, gauche boy through an adolescence in which
his mathematical ability began to blossom, to the achievements of his maturity,
the story of Turing's life fascinates. In the years since his suicide, Turing's
reputation has only grown, as his contributions to logic, mathematics,
computing, artificial intelligence and computational biology have
become better appreciated.

To commemorate the centenary of Turing's birth, this republication of his
mother's biography, unavailable for many years, is enriched by a new foreword
by Martin Davis and a never-before-published memoir by Alan's older
brother. The contrast between this memoir and the original biography reveals
tensions and sheds new light on Turing's relationship with his family,
and on the man himself.

Alan M.
TURING

Centenary Edition

Sara TURING

with a foreword by
Martin Davis

with an Afterword by
John Turing

CAMBRIDGE
UNIVERSITY PRESS

University Printing House, Cambridge CB2 8BS, United Kingdom

Cambridge University Press is part of the University of Cambridge.

It furthers the University's mission by disseminating knowledge in the pursuit of education, learning and research at the highest international levels of excellence.

www.cambridge.org
Information on this title: www.cambridge.org/9781107020580

© S. Turing 1959, 2012

First edition published by W. Heffer & Sons, Ltd 1959
Second edition published 2012
5th printing 2014

Foreword to the First Edition © L. Irvine 1959, 2012
Foreword to the Second Edition © M. Davis 2012
Afterword © J. Turing 2012

Printed in the United Kingdom by CPI Group Ltd, Croydon CR0 4YY

A catalogue record for this publication is available from the British Library

ISBN 978-1-107-02058-0 Hardback

CONTENTS

≈

Part Two • Concerning Computing Machinery
and Morphogenesis

FOREWORD
TO THE CENTENARY EDITION

≈

S ara Turing, a woman in her seventies mourning the death of
Alan, her younger son, a man that she failed to understand
on so many levels, wrote this remarkable biographical essay. She
carefully pieced together his school reports, copies of his publications, and comments on his achievements by experts. But Alan
Turing was a thoroughly unconventional man, whose method of
dealing with life's situations was to think everything through from
first principles, ignoring social expectations. And she was trying
to fit him into a framework that reveals more about her and her
social situation than it does about him. Alan's older brother John
trying to fill in the gaps he saw in his mother's account, also ends
up revealing a good deal about his own attitudes.In these few pages
I will discuss some of the questions that may occur to readers of
these documents.

Alan Turing's War

In 1940, after France had been defeated, Britain fought on mainly
alone. The merchant shipping on which the island was dependent
was being sunk by German submarines at a rate that threatened
to force the UK to yield. The radio communications between the
submarines and their base concerning their operational plans were
being picked up in Britain. If these plans were known, attacks could
be mounted against the submarines. Merchant ships could adjust
their routes so as not to go where they would encounter enemy
submarines. But of course the data was encrypted.

For the purpose of decrypting enemy communications, an assorted group of classics scholars, mathematicians, and hobbyists who were good at solving puzzles were brought together in an estate called Bletchley Park near the present-day town of Milton Keynes. For much military communication the Germans used an enhanced version of a commercial enciphering machine called Enigma. Some Polish mathematicians had worked out a technique for decrypting German military messages coded on their Enigma, and had passed their work on to the English. But by the time war broke out in 1939, the Germans had added additional complications and the Polish techniques were no longer of any use. Not only were there several disks on the machine whose rotational position could be altered, but there was also a plug-board into which cables could be plugged in various ways. In order to decrypt a message, these precise settings had to be known. Taking advantage of certain weaknesses in the design of the machines as well as carelessness on the part of German cipher clerks, the Bletchley park code-breakers could rule out a large number of possible settings. That still left a number of possibilities to be attacked by trial and error. Turing played a major role in developing these techniques, and in working out a method for automating them. He designed a machine that would systematically try various settings, rejecting those that contradicted what was already known. These machines, many of which were built, strangely called bombes, were highly effective. What is truly remarkable is that, constructed to Turing's specifications, they worked as intended without the need for any fine tuning. Although Turing's contributions, and indeed the entire project of decrypting German military communications, were kept secret long after the war, Turing was awarded an O.B.E. (Order of the British Empire) for his contributions to the war effort.

Alan Turing's Universal Computer

Mathematical proofs use logical reasoning to get from assertions already accepted as true to statements called theorems, which thus achieve acceptance as mathematical truths. The work of logicians in the nineteenth and twentieth centuries showed how, in principle, the individual steps in such "proofs" could be replaced by the mechanical manipulation of symbols. This situation gave rise to the problem of finding a mechanical process, an algorithm, for deciding in advance whether from some given statements accepted as true, another desired statement could be obtained by such a sequence of steps. The great mathematician David Hilbert declared that this problem, he called the *Entscheidungsproblem*, was the main problem of mathematical logic. (The long German word simply translates into "decision problem," but since many problems involve "decisions," it has been customary to use the German name.) The game of chess provides a useful analogy. The individual moves in a chess game, like the individual steps in a logical proof, are simple and mechanical. The *Entscheidungsproblem* is then like the problem of how to tell for a given initial position of the chess pieces, whether white can achieve a check-mate regardless of black's counter moves. As every chess player knows, this is very difficult if not impossible.

Alan Turing learned about the *Entscheidungsproblem* from lectures on the foundations of mathematics given by Max Newman at Cambridge University in 1935. People were not at all convinced that there could be an algorithm meeting Hilbert's requirements. The mathematician G.H. Hardy, a professor at Cambridge, said it forcefully:

> There is of course no such [algorithm], and this is very fortunate, since if there were we should have a mechanical set of rules for the solution of all mathematical problems, and our activities as mathematicians would come to an end.

Alan agreed, and considered how one could go about proving that no such algorithm exists. Apparently no one had ever provided a definition of "algorithm," and indeed there had hardly been a need for such a definition. As children, we all learned algorithms for adding and multiplying numbers. Later many of us will have learned algorithms for solving equations, and even the algorithms from the differential calculus for computing derivatives. None of this required that we be told what an algorithm is. We recognized that the rules we used were explicit and mechanical. Once we learned them, we could carry them out without any creative thought. That was good enough. But to prove that there is no algorithm to carry out some task, more was needed than the words "explicit" and "mechanical."

Thinking about what people do when they "compute," that is carry out algorithms, Turing saw that it all seemed to amount to taking note of particular symbols and then writing other symbols. Although the work is done on a two-dimensional surface like a sheet of paper or a blackboard, Turing could see that, in principle, it could all be done on a paper tape in which the symbols are written as a linear string. He realized that it was crucial that no limit be placed in advance on the amount of space needed. Speaking somewhat metaphorically, the tape should be infinitely long to be sure that it doesn't get all used up before the computation is complete. Next he saw that the behavior of the person doing the computation could be represented by a simple table that indicated the next step to be carried out, writing a symbol and moving left or right on the tape. Finally a machine could be constructed that does what the table instructs it to do. Such machines have come to be called Turing machines.

Now he was off and running. He asked whether one of his machines could be what we may call a *tester*. What a tester would be required to do would be to determine whether a given one of his machines will eventually write some particular symbol, say "0", when started with its tape empty. One imagines writing the table for the machine being tested on the tape of the "tester" and expecting it to eventually halt with "yes" written on its tape if the machine being tested would eventually write "0," and "no" written on its tape, otherwise. Turing proved that no such tester could exist. Finally, he showed how to use symbolic logic to represent the behavior of his machines in such a way that an algorithm to solve the *Entscheidungsproblem* could also be used as a tester. The conclusion: Since there are no testers, there also is no algorithm to solve the *Entscheidungsproblem*.

This work was enough for a very important research paper. Turing had given an explicit characterization of what is algorithmically computable, had provided a simple example of a problem that is not algorithmically solvable, and had used this to prove that the *Entscheidungsproblem* itself is unsolvable. Indeed, Turing's paper "On Computable Numbers with an Application to the Entscheidungsproblem" published in 1936 did all of that. But it did something more, something that was to apply not only to an abstract mathematical problem, but also to a matter of great practical importance: the possibility of making an all-purpose machine for computation, a machine that could crunch numbers to work out how to get to the moon, but could also play a game of chess, as well as carry out the many other tasks we have learned to entrust to what we now call "computers." Turing wrote out in detail the table for a machine U that he called "universal." What made it deserve this name is that if the table for one of his machines M is written on U's tape, and U is then started, it will carry out the same calculation that M would do when started on a blank tape. So U can do anything that is computable. Of course due to limitations of space and time, no physical device can be fully universal. But what Turing's paper made clear is that given the capability to do a few simple things together with a very large memory, a physical machine could approximate universality.

It's important to emphasize that although the "machines" in Turing's 1936 paper existed, and were only intended to exist, on paper, nevertheless they represented a paradigm shift in the way people thought about computation. Computation was not only "number crunching," but was also the execution of algorithms dealing with any kind of data. Moreover, they showed that the distinction between what would come to be called hardware and software, as well as that between program and data were quite relative to circumstance and convenience. Turing's machines were conceived as machines, but their tables on the tape of the universal machine functioned as programs, and the universal machine treated them as data. When the question of how much functionality to build into the hardware of a computer became a practical one, these considerations dominated the discussion. John von Neumann, who wanted his machine to do number crunching in connection with the design of hydrogen bombs, chose to make basic arithmetic part of the hardware. Turing, who since his Bletchley Park days, had imagined a computer capable of playing decent chess, chose in his design to build only fundamental logical operations into the hardware, with arithmetic to be supplied by programming. Turing was dismissive of the "American tradition of solving one's difficulties by means of much equipment rather than by thought." Many years later, when the technology had advanced to the point that the universal computer was embedded on a chip of silicon and the huge memory was just more silicon, this issue was debated in terms of RISC (Reduced Instruction Set Computing) vs. CISC (Complete Instruction Set Computing) computer architecture.

Alan Turing's Homosexuality

Turing had been a practicing homosexual since his puberty, and apparently regarded his sexuality as simply part of who he was. It is not clear what his mother knew of this, but when his arrest for engaging in a liaison with a young man in Manchester made it a matter of public record, he did provide her with some explanation. In any case, her biography says nothing about any of this. His brother's essay does discuss it, and even offers an explanation of his brother's orientation. He blames it on Alan having been left to board with strangers in England at a very young age when the needs of the British Raj called his parents to India. Such psychological explanations of homosexuality, considered a "disorder," were very much the vogue at the time when John Turing was writing. Alan himself angrily suggested something similar on an occasion when one of his advances had been rebuffed.

Nowadays, it is accepted that the specific sexual proclivities of an individual are simply one aspect of that person and are unlikely to be altered by any intervention. There is apparently some evidence to connect male homosexuality to hormonal influences in the womb. But the truth is that, at this time, it is just not understood why certain people are homosexual. In any case, if it is reasonable to want to understand why a specific individual has turned out to be homosexual, it is just as reasonable to seek to understand why another is heterosexual. The powerful force of sexual attraction remains deeply mysterious.

The Engagement

Although Sara Turing doesn't mention it, Alan was engaged to a young woman, Joan Clarke, a co-worker at Bletchley Park, for several months. Joan was a very talented mathematics student who had been recruited to work as a cryptanalyst. Although Alan let her know from the beginning of his homosexual "tendencies," she remained willing to continue the engagement. It was after they spent a week together on a bicycling trip in Wales, that he decided that it wouldn't work, and broke off the engagement. They were, and remained, very fond of one another and it was all very difficult. Many years later, she decided not to see the play *Breaking the Code* which was about Turing, because she would have found it too painful.

Joan dealt with institutional obstacles and social prejudices that faced anyone who was both female and a mathematician at that time. At Bletchley Park she was listed as a "linguist" because the designation "cryptanalyst" was not available to women. She later married a man with a great interest in Scottish history. She became similarly interested and made a significant contribution to numismatics in that connection. She died in 1996.

John Turing dismissed Joan as "safe" (apparently meaning unattractive). In an earlier version of the document, he had referred to her "unwashed hair" and "problems of personal hygiene." This was in contrast with the "attractive and lively young women" that John had brought home for weekends who "cheered up" his father. Even allowing for the prejudices of the time, this denigration of a capable intelligent woman was truly appalling.

Turing in Princeton

Although Alan had no way of knowing this, he had not been the only one working on the problem of characterizing algorithmic computability. In Princeton Kurt Gödel visiting from Vienna as well as Alonzo Church and his students at Princeton University discussed the same problem. This came to light at Cambridge when a copy of a mathematical periodical arrived in the mail containing an article by Church with the title "An Unsolvable Problem of Elementary Number Theory." It turned out that, in addition, Church had also published a proof of the unsolvability of the *Entscheidungsproblem*. So, in a sense, Alan had been preempted. But his approach was so different and so fundamental that it was clearly still important enough to merit publication. Also the notion of universality, with its implied consequences for a new understanding of the nature of computation, was entirely Turing's. It was decided that Alan should spend some time at Princeton so Church and he could explore their common interests, and Max Newman wrote Church to see what could be worked out. Turing did spend two years at Princeton followed by a year back in Cambridge before the war broke out. There was something anomalous about Alan's situation at Princeton. At that time, in England, a doctorate was not ordinarily considered part of the preparation for an academic career. As a Fellow, Turing was on the lower rungs of a ladder that, if his research career proved successful, could eventually lead to a professorship. But the American system was different, and the simplest way for Turing to fit himself into it was as a candidate for the Ph.D. His dissertation was an important contribution to mathematical logic extending Gödel's work on undecidability.

Turing's ACE

After the war had ended, Alan was eager to help build a working Universal computer. He was offered a position at the National Physical Laboratory (NPL) to do just that. Full of enthusiasm, and harnessing the practical knowledge of electronics he had acquired from his war work, he wrote a detailed plan for a machine he called Automatic Computing Engine (ACE). This document and the machine it proposed anticipated a number of concepts that later were widely accepted. An address Turing delivered to the London Mathematical Society on the proposed ACE demonstrated Turing's expansive view of what came to be called computer science.

Unfortunately, the project ran into bureaucratic difficulties Turing had not expected, being used to the war-time atmosphere in which obstacles of that nature could be eliminated by a letter to Winston Churchill. In addition, engineers ignorant of the great success of Turing's "bombe" at Bletchley Park, wouldn't take seriously the pronouncements of this stuttering mathematician. He must have been terribly frustrated when computers did come to be built elsewhere, and their design moreover followed "the Americans" in solving problems by hardware rather than "thought." He left and accepted Max Newman's invitation to come to Manchester to work with the computer being built there. He did, but interacted with it not in pushing the kind of advanced software development he had outlined in his ACE report, but rather as a user to carry out computations related to the biological problems in which he had become interested.

Can Machines Think?

What led Turing to raise this question was that he saw in his ACE a first crude approximation to a human brain. He wrote a much-cited essay on the subject and even spoke about it on the radio. He sought an objective experimental test on the basis of which one could be justified in saying that a programmed computer was thinking, avoiding philosophical and religious objections that might be raised. The criterion he chose was the ability of such a machine to carry on a conversation that could not reasonably be distinguished from one by a person. He predicted that this would be achieved by the end of the twentieth century, but was far too optimistic about the task of programming computers to achieve a command of natural language equivalent to that of every normal person.

The Burglary

Alan foolishly went to the police when a few items had been stolen from his house. Alan's sex partner Arnold had mentioned Turing's posh house to someone named Harry, and Harry had gone to the house and helped himself. It turned out that Harry had been known to the police and had left his fingerprints behind after the theft. John Turing thought that Arnold himself was the thief and that there had been no burglary, but he was pretty clearly mistaken.

In any case, the police were more interested in what Alan and Arnold had been doing together than in the theft, and Turing found himself before a judge charged with "gross indecency." In order to spare Turing from a prison sentence, he was required to undergo a course of estrogen injections for a year, apparently in an effort to block his sex drive. What it did accomplish was to cause Alan to grow breasts.

Alan Turing's Death

Sara Turing would have it that it was her son's slovenly habits that led to his getting deadly cyanide on an apple he was eating. John

Turing was convinced that it was suicide. Alan Turing was a man who was privy to official secrets that, after his conviction, he was no longer entitled to have. Sex for him in England was evidently dangerous. In the Cold War atmosphere of the 1950s, he would surely have been warned about travel abroad. When a man he had met in Norway tried to visit Turing, the authorities saw to it that it wouldn't happen.

In any case, there is reason to believe that Alan did take his life, and that moreover he had staged his suicide in such a way that it would be clear to friends what he had done, while to his mother it would appear as a vindication of all her warnings about his slovenly habits. He had been much impressed by the Walt Disney film *Snow White and the Seven Dwarfs* and particularly the scene in which the wicked witch holds an apple in a steaming pot of poison chanting:

> Dip the apple in the brew
> Let the Sleeping Death seep through.

We are told that Alan enjoyed chanting those lines. Perhaps this was the very song he sang as he prepared the deadly concoction and took his bite.

Other Reading

First and foremost, there is Andrew Hodges's biographical master-piece, *Alan Turing: The Enigma*. A much shorter and very worth-while account is David Leavitt's *The Man Who Knew too Much: Alan Turing and the Invention of the Computer*. Finally I venture to mention my own *The Universal Computer: The Path from Leibniz to Turing* which tells the story of developments leading up to Turing's breakthrough, now available in an updated edition for the Turing Centenary.

Martin Davis

PREFACE TO THE FIRST EDITION

≋

The aim of this book is to trace from early days the development of a mathematician and scientist of great originality and to record details from which a selection may be made by a future biographer. Owing to the enforced silence regarding my son's activities in the Foreign Office during the Second World War, there is, except for some few anecdotes, a regrettable gap of six years in the narrative. The book is divided into two parts. The former and major part is mainly biographical but contains sufficient scientific material to indicate the scope and depth of my son's research. The second part affords more technical particulars which might prove wearisome to the general reader: but these barely touch the fringe of his work on Computing Machines and Morphogenesis. His writings on these and other subjects, together with a posthumous paper on Morphogenesis, prepared by Dr. N.E. Hoskin and Dr. B. Richards, can be studied in the volume of his collected works, which is to be published by the North-Holland Publishing Company, Amsterdam. I am indebted to many of my son's friends – too numerous to name – for their recollections. Here I take this opportunity to express my gratitude to various American mathematicians and scientists for their particularly courteous interest and co-operation. For their scrutiny of my typescript and valuable suggestions and advice, my special thanks go to Professor M.H.A. Newman, F.R.S., Mrs. Newman (Lyn Irvine), Mr. Geoffrey O'Hanlon and to Mr. Nowell Smith, who also read the proofs.

E.S.T.

FOREWORD TO THE FIRST EDITION

≋

This book contains almost all the essential material for the biography of a very remarkable man, who died tragically in June, 1954, in the prime of his life and in the middle of research which may still prove to be even more original and important than the finished work which had brought him so much honour and fame. Alan Turing's mother, who has assembled and written this record of his childhood and his mature achievements, believes that his death was accidental. The explanation of suicide will never satisfy those who were in close touch with Alan during the last months and days of his life, however much the available evidence may point to it, and in the future the possibility of accident will be considered by those in a better position perhaps to decide the truth. But even if his death was not chosen by him, he was a very strange man, one who never fitted in anywhere quite successfully. His scattered efforts to appear at home in the upper middle class circles into which he was born stand out as particularly unsuccessful. He did adopt a few conventions, apparently at random, but he discarded the majority of their ways and ideas without hesitation or apology. Unfortunately the ways of the academic world which might have proved his refuge, puzzled and bored him; and in return that world sometimes accepted him wholeheartedly (I remember Shaun Wylie's saying "He was a lovely man: never a dull moment") but often felt puzzled by his remoteness. A letter from Sir Geoffrey Jefferson to Sara Turing describes this particularly well:

He was a wonderful chap in many ways. I remember how he came to my house late one evening to talk to Professor J.Z. Young and me after we had been to a meeting in the Philosophy Department here, arranged by Professor Emmet. I was worried about him because he had come hungry through the rain on his cycle with nothing but an inadequate cape and no hat. After midnight he went off to ride home some five miles or so through the same winter's rain. He thought so little of the physical discomfort that he did not seem to apprehend in the least degree why we felt concerned about him, and refused all help. It was as if he lived in a different and (I add diffidently, my impression) slightly inhuman world. Yet he had some warmth, I know – for you in particular, for he told me so in a revealing couple of hours that we had together not very long before he died. . . . Alan, as I saw him, made people want to help and protect him though he was rather insulated from human relations. Or perhaps because of that we wanted to break through. I personally did not find him easy to get close to.

We all marvelled at his indifference to creature comforts – for example, his staying at YMCA Hostels when he could easily afford a first class hotel. But was he so indifferent? He always appreciated finding himself warm and well-fed in a strange house during the difficult winters after the war. But he was at least half a Spartan and did not believe in expending much trouble and expense on physical comforts. He was Spartan rather than Bohemian. At Bletchley during the war when crockery was scarce and expensive, it was a nuisance if one's tea mug disappeared and Alan with characteristic thoroughness brought a padlock and chain and locked his mug to the radiator in his room. He was genuinely furious when some wag took the trouble to pick the lock and hide the mug.

Alan certainly had less of the eighteenth and nineteenth centuries in him than most of his contemporaries. One must go back three centuries (or *on* two perhaps) to place him; and yet of all the great minds most likely to understand and appreciate him, I should place Tolstoy first. A couple of years before he died I pushed first *Anna Karenina* and then *War and Peace* into his hands. I knew that he read Jane Austen and Trollope as sedatives, but he was totally uninterested in poetry and not particularly sensitive to literature or any of the arts, and therefore not at all an easy person to supply with reading matter. *War and Peace* proved to be in a very special way the masterpiece for him and he wrote to me expressing in moving terms his appreciation of Tolstoy's understanding and insight. Alan had recognized himself and his own problems in *War and Peace* and Tolstoy had gained a new reader of a moral stature and complexity and an originality of spirit equal to his own.

With ninety-nine people out of a hundred Alan protected himself by his off-hand manners and his long silences – silences finally torn up by the shrill stammer and the crowing laugh which told upon the nerves even of his friends. He had a strange way of not meeting the eye, of sidling out of the door with a brusque and off-hand word of thanks. His oddly-contoured head, handsome and even imposing, suddenly from another angle, or in a different mood, became unprepossessing. He never looked right in his clothes, neither in his Burberry, well-worn, dirty, and a size too small, nor when he took pains and wore a clean white shirt or his best blue tweed suit. An Alchemist's robe, or chain mail would have suited him, the first one fitting in with his abstracted manner, the second with that dark powerful head, with its chin like a ship's prow and its nose short and curved like the nose of an enquiring animal. The chain mail would have gone with his eyes too, blue to the brightness and richness of stained glass. They sometimes passed unnoticed at first; he had a way of keeping them to himself, and there was also so much that was curious and interesting about his appearance to distract the attention. But once he had looked directly and earnestly at his companion, in the confidence of friendly talk, his eyes could never again be missed. Such candour and comprehension looked from them, something so civilized that one hardly dared to breathe. Being so far beyond words and acts, that glance seemed also beyond humanity.

It was more than fortunate for Alan that his mother took such pains to select a public school to suit him. I find the account of his years at Sherborne fascinating. His mother and his housemaster, of one mind about him throughout, saved Alan from what threatened to be a career of scientific pranks. It was through Sara Turing's appreciation at a very early stage both of his brilliance and his difficulties, that he went to Sherborne and went in good heart, and there in a housemaster of unique perception and tact, he found someone able to carry on the difficult task of discouraging the misfit without discouraging the genius. It was at Sherborne, in his deep attachment to Christopher Morcom, the brilliant boy who died at eighteen, that Alan saw a vision of human relationships which sent him questing for the rest of his life.

To those who like myself came to know Alan only in the last ten years of his life, there is the answer to many questions in this short book. He carried so many odd suggestions of his past as well as his present about with him, almost like a pedlar's wares festooned about his person, although without any notion of showing them off. From his being in part still a child and an adolescent and an undergraduate, as well as a don and a Fellow of the Royal Society, arose the extraordinarily wide range of his friends. No one I have ever known proved compatible with so many people who would themselves have been incompatible with one another; partly through his divine tardiness to notice the faults of anyone who had won his regard, no matter by what trifling service. Yet it was characteristic of his honesty and detachment that he would listen to criticisms of his friends with the same humility as he accepted criticisms of himself – never apparently suspecting that people can find fault except honestly and from the best motives. He himself found the idea of deceiving others so distasteful that he supposed it equally so to almost everyone.

It is hard to remember a single instance of Alan's acting in imitation, even unconscious imitation, of another person. His originality was something quite by itself in its extent and depth. Sara

Turing quotes his writing to her as a little boy from his preparatory school saying: "I seem always to want to make things from the thing that is commonest in nature." This throughout was his ruling principle, and some of the ways that it guided and affected his research are described in this book. The specialist will be able to trace it in many of his more. important interests. To others it was particularly evident in his long-distance running. In that he achieved by mere legs and feet what most of us achieve only with the help of horses or wheels and the internal combustion engine. In the Easter holidays of 1949 Alan stayed with us[1] at Criccieth (the Pearsons having lent us their charming house in Marine Terrace). One afternoon of overcast skies and threatened rain, Alan changed into blue shorts and disappeared for a short time. When we asked him where he had been he pointed out a promontory of Cardigan Bay seven or eight miles north-west, inaccessible by road. We might have entertained the idea of walking there, but not without carrying a meal and macintoshes with us, scarcely without resting an hour or so on the way. For us it would have been a day's outing, but Alan did it between lunch and tea. From that day-although his normal walking gait was uninspired and almost shambling-we all felt awed, as if Mercury had joined our circle of acquaintances.

That sentence from his childish letter without other evidence would stamp him as a genius. He was not merely doing something extraordinary in a small boy but *recognizing it as extraordinary*, and it is the recognition of self that carries genius through to achievement. I recommend this record to anyone who has an interest in the nature of what we call genius. We are very ignorant still about its origin and character and it is hard to see how this ignorance can be readily mended, owing to the lack of material to study. There are very few men and women of genius in any century, and of these few some are certain to be overlooked in their own time and possibly in all time. Of those who are known the material for biography

---

[1] Professor and Mrs M.H.A. Newman.

is often thin and dull, and particularly with men of science much has to be made of little. Sir Isaac Newton remains a mystery to us after every recorded morsel has been displayed and examined. A new anecdote of his youth would be seized as a treasure by the entire learned world, no matter how slight and trivial an anecdote it might be, for it is particularly the child and the boy that excites most curiosity. We want to know where and how a nature and mind so unlike the normal first showed its divergence.

Sara Turing survives her son owing to the tragic earliness of his death, and with great courage and faith she has taken the opportunity that his death offered and made this source-book for a future biographer. For it does show unusual courage not to be ashamed of putting down the trifling memories, the details of childhood and family affairs, the little events that are almost insignificant and yet have just that faint signature in the corner, "A.M.T.," which made them so well worth preserving. Nothing that science can ever offer is more valuable than the knowledge of how a scientist develops.

Lyn Irvine.

PART ONE

Mainly Biographical

1

Family Background

The Turing family is of Norman extraction and the family tree goes back to 1316 AD, the family motto being *Fortuna audentes Juvat*. Having arrived in Scotland the members settled in Angus in a barony of that name, whence they removed to Aberdeenshire early in the fourteenth century and came into possession of Foveran, which remained the family seat until recent times. The name was variously spelled Turyne, Thuring, Turin, Turing. William Turin received the honour of knighthood from James VI of Scotland (James I of England) and thereafter Sir William added the final "g" to the name.

John Turing of Foveran was created a baronet by Charles I in 1639 for loyal service, and was at the battle of Worcester; but his loyalty cost him the loss of lands which had been in the family for 300 years. Records show Turings holding positions of trust and responsibility in the County of Aberdeen.

By the eighteenth century some Turings were venturing further afield. Thus Sir Robert Turing (Bart.), born in 1744, was a doctor and amassed a considerable fortune in the East Indies and then retired to Banff in Scotland where he made himself very useful and popular. One kinsman in the Honourable East India Company took part in the defence of Seringapatam. Others in the nineteenth century lived in Holland; two, father and son, were successive British Consuls in Rotterdam. Some of their descendants have now become domiciled in Holland. Alan's great grandfather, presumably through this Dutch connection, had some occupation in Batavia, maybe in some shipping concern. He was John Robert

Turing (1793–1828) who married Jane S. Fraser, and it was, I think, on a voyage back from Batavia that his family were involved in a shipwreck.

His son, another John Robert Turing, who was Alan's grandfather, was admitted to Trinity College, Cambridge, in May, 1844, and in the Mathematical Tripos 1848 was classed eleventh among the "Senior optimes." At Trinity he was notorious for sleep-walking on the leads. In 1848 he was ordained Deacon, and Priest in 1849, and was Chaplain of Trinity College from 1859 to 1871, and simultaneously from 1859 to 1864 was Curate at Great St. Mary's, Cambridge. Marrying Fanny Montagu Boyd he had ten children of whom eight survived. It was when he was Rector of Edwinstowe, Nottingham, that his son, Alan's father, Julius Mathison Turing, was born, 9th November, 1873. On the death of his father, when Julius was ten, the family moved to Bedford: later from Bedford School Julius won a history scholarship to Corpus Christi College, Oxford, and thence passed into the Indian Civil Service, and was posted to the Madras Presidency. He inherited none of his father's mathematical ability, in fact algebra was just mumbo-jumbo to him and as for the claimed result of one minus quantity multiplied by another minus quantity – that for him was beyond human comprehension.

On the maternal side Alan was descended from the Stoneys. According to Burke's Landed Gentry of Ireland the Stoneys are believed to be descended from a Danish family which settled near Kettlewell in Craven in Yorkshire about the ninth century and were known by the name 'de Stanehow,' or 'Stonehow.' One member of the family was Rector of Kettlewell about the time of Edward I and others were among those who paid Richard II's Poll Tax in 1379 at Buckden, three miles from Kettlewell. Sundry domestic events are recorded in the register of the Church at Rilston, Yorkshire, among these the marriage on 6th January, 1675 (date according to "old style" – by our reckoning 1676), of George Stoney of Kettlewell and

Mary Moorhouse of Rilston, direct ancestors of the Irish Stoneys.[1] George and Mary Stoney emigrated to Southern Ireland at the end of the seventeenth century obtaining land under the William and Mary scheme which offered inducements to English Protestants with capital to settle there. George Stoney took up his abode at Knockshegowna (Hill of the Fairies) in the northern extremity of Tipperary. Alan's great-great-great-great-uncle, Andrew Robinson Stoney, subsequently known as Bowes, married the Dowager Countess of Strathmore; under her father's will any person whom she married had to assume her maiden name to assure her inheritance. Hence the coupling of the name Bowes with the family name, Lyon, of the Earls of Strathmore. It is an understatement to add that Andrew Stoney Bowes was no adornment to the family.

The *Annals of the Stoney Family* show its members leading the ordinary life of the "landed gentry" in County Tipperary and King's County, occasionally sending sons to England for education and occupied with the supervision of their estates and livestock and with hunting. Some held positions of responsibility as J.P.s and so forth, one being Deputy Governor of Tipperary. There is something pleasantly feudal in the account of my father's great-grandfather, the principal magistrate in the neighbourhood usually holding a petty sessions court on his front door steps, while an arm-chair in the porch served as the "bench." Sundays saw large dinner parties of twenty to thirty guests at his home, Arran Hill, to which relations and intimate friends had standing invitations. He always enjoyed showing visitors his deer park and herd of Devon cattle. Open house was kept: guests stayed as long as they liked to hunt with their host's private hounds. In the little church at Borrisokane the Stoney's pew was a small room off the chancel with its own open fire – all very snug. It was the privilege of the eldest son to occupy a comer seat whence he could survey the congregation. This room

[1] Here I follow the record given in *Annals of the Stoney Family* by Major F. S. Stoney, R.A.

is now put to another use and houses the stove to heat the church, but a tablet above the door commemorates its having been the "Stoney Pew."

Thomas George Stoney, J.P., of Kyle Park, Co. Tipperary, Alan's maternal great-grandfather, married in 1829, Anna Henrietta Waller, a member of the family of Wallers, among whom were Sir William Waller (known by Londoners as "William the Conqueror"), a highly skilled General in Cromwell's army, and his first cousin, Hardress Waller. The latter was one of the Regicide Judges; but in 1660 he professed his penitence, adding that he "did appear more to preserve the King upon trial and sentence than any other." His petition for pardon is among the Egerton manuscripts in the British Museum.

This Thomas George Stoney (my grandfather) was a man of considerable enterprise. Over a hundred years ago he introduced on his lands mechanical reapers which had to be conveyed about sixty miles from the nearest railway. I have seen both a model of the school which he intended for the children of his employees and a specimen of the £1 notes signed and issued by him for use on his estates. However he "wasted his substance" on building and horses; so two of his sons, Francis G.M. Stoney and Edward Waller Stoney (my father), became civil engineers. The former, Alan's great-uncle, invented the "Stoney Sluice" used on the Assuan Dam, the Manchester Ship Canal, in the bridge over the Thames at Richmond and at numerous other places the world over; he was also the inventor of the "Titan Cranes." The story is told that Francis Stoney, on going up to be interviewed for a certain post, took a model of his sluice, the working of which he demonstrated to the other waiting candidates. Francis was the first to be summoned to the interview – when he emerged he found the waiting room empty; the other candidates, completely discouraged, had disappeared.

Edward Waller Stoney, C.I.E., Alan's maternal grandfather, spent most of his professional life as an engineer of the Madras and Southern Mahratta Railway, of which he later became Chief Engineer.

His inventiveness came out in the original methods he devised for the construction of bridges over some of the great Indian rivers, notably the Tangabudra. In connection with railways he brought out various patents, but to Anglo-Indians he was best known as the inventor of "Stoney's Patent Silent Punkah-wheel." Previously sleep was much disturbed by the creaking punkah- wheels. In 1903 he was made a Companion of the Indian Empire.

A distinguished collateral relation of Alan's was Dr. George Johnstone Stoney, F.R.S., president of the Royal Dublin Society. He pre-supposed the existence of the electron and in its hypothetical stage named it and also gave the name to ultra-violet rays. He was known as "Electron Stoney": with all his learning he used to say, "we know so little." He was one of the great who nevertheless retained into old age a childlike simplicity. I well remember him in his eighties with his long, flowing, snow-white beard. Age had not dimmed his enthusiasms. Gramophones were a new invention and he owned the most enormous one, from which he derived great pleasure. His knowledge of music was such that he used to compose himself for sleep by reading musical scores. Another great interest of his old age was the study of Esperanto, which he believed had a great future. His son, George Gerald Stoney, F.R.S., made his name in connection with work on steam turbines in collaboration with Charles, later Sir Charles, Parsons. From his father he had learned much about the technique of silvering of mirrors, which led to his appointment in 1893 as Manager, in addition to other duties, of the Searchlight Reflector Department of Messrs. C.A. Parsons & Co. Though he had resigned from the firm in 1912 he gave much advice on the re-organization of the searchlight mirror department which in the 1914–18 war had become the largest of its kind in the world. He likewise served on Lord Fisher's board of invention and research, and later on the Lancashire anti-submarine committee. (These activities are interesting to compare with Alan's work for the Foreign Office in the Second World War.)

Johnstone Stoney's eldest daughter, Dr. Florence Stoney, in her hospital in France was very early in the field using X-ray photography to locate shrapnel and bullets in the wounded in the 1914–18 war. Her younger sister, Edith A. Stoney, who had been at Newnham College, Cambridge, and in the early nineties was bracketed with the seventeenth wrangler, did remarkable service in the 1914–18 war with the Scottish Women's Unit in Serbia, setting up and working electrical equipment for their hospital. Bindon Blood Stoney, F.R.S., brother of Johnstone, was noted for his invention of his "Shears Float" for dredging Dublin Harbour and of docks which may be regarded as the forerunners of the famous Mulberry Dock.

Alan's maternal grandmother was Sarah Crawford of Cartron Abbey, Longford, Ireland. It was her grandmother, (I think), a Miss Lindsay, belonging to the family of the Earl of Crawford and Balcarres, who married a Mr. Crawford. My mother recalled how the old family servants used to speak of the bride arriving provided with house linen all embroidered with a coronet. These Crawfords (or Crawfurd as originally spelt) were related to the Crawfurds of Crawfurd – John, through the houses of Loudoun and Kilbirny. My mother was a gifted amateur painter and excelled in her beautiful paintings of the wild flowers of the Nilgiri Hills in South India. A volume of these flower paintings was presented by me and most gratefully accepted by the Royal Botanical Gardens, Kew, where, to quote the Director, "it forms a valuable addition to the collection of drawings."

I was Ethel Sara, daughter of Edward Waller Stoney, and was born at Podanur, Madras Presidency, on 18th November, 1881. My education was at the Alexandra School and College, Dublin, and at Cheltenham Ladies' College. Later I attended lectures at the Sorbonne in Paris, before I joined my parents in Madras.

On a voyage home from India via Japan, Canada and the United States of America I met Julius Turing and we were married in Dublin on 1st October, 1907.

2

Childhood and Early Boyhood

S hortly after our marriage my husband and I returned to the
Madras Presidency. Our elder son, John Ferrier, was born at
my parents' home in Coonoor, Nilgiris, 1st September, 1908. Nearly
four years later our second son, Alan Mathison, was born at Warrington Lodge, Warrington Crescent, Maida Vale, London, on 23rd
June, 1912. His christening took place at St. Saviour's Church,
Warrington Avenue, on 7th July, 1912. We spent the following
winter with our boys in Italy. My husband returned to India in the
spring of 1913, while I followed in September, leaving both children at home with Colonel and Mrs. Ward at St. Leonards-on-Sea.
Both boys grew very much attached to "Grannie" as they called
Mrs. Ward. It had been intended to take Alan out to India, but
owing to his having slight ricketts it was thought better to leave
him in England. Despite his delicacy he was an extremely vivacious
and forthcoming small child.

My letters to my husband when I was in England in the spring
and summer of 1915, round about Alan's third birthday, give some
idea of what he was like. I was not alone in my opinion when I
wrote, "a very clever child, I should say, with a wonderful memory
for new words," for I reported that his uncle, Herbert Trustram
Eve, maintained that he would do great things – this when Alan
was nearly three. Here are extracts from letters at this time: "Alan
generally speaks remarkably correctly and well. He has rather a
delightful phrase, 'for so many morrows,' which we think means,
'for a long time,' and is used with reference to past or future." Being
a very pretty and engaging small boy he attracted a good deal of

9

notice from complete strangers, and workmen who came to the house. In those days he was quite free from shyness and ready to greet anyone. He loved hunting up in John's history book the picture of what he called "the firstest train" (Stephenson's). One letter to my husband in May, 1915, says: "Alan will in a moment cry with rage and attempt to hold his breath, and in the next moment he will laugh at his tears, saying 'Look at my big tears,' and squeeze his eyes and say, 'Ah' with his mouth wide open trying to squeeze out more tears for fun." However, three months later a letter says, "Alan has improved greatly. He has many charming traits. He misses nothing. The maid in these rooms took my newspaper without leave . . . when she was taking the tea away Alan ran off saying, 'I'll get the paper for Elsie (the maid) to read,' which though useful was not tactful." In this summer of 1915 he made his first venture in experimenting: as one of the wooden sailors in his toy boat had got broken he planted the arms and legs in the garden, confident that they would grow into toy sailors.

Once more in the autumn of 1915 I sailed for India and the children were left at St. Leonards-on-Sea pending the return of my husband and myself in the spring of 1916. On the expiry of his leave my husband returned in the late autumn to India but did not wish me to have a fourth voyage among enemy submarines, so I had the boys with me in "rooms" at St. Leonards-on-Sea until the end of the war, with John home only for the holidays.

Alan was interested in figures – not with any mathematical association – before he could read and would study the numbers on lamp posts, etc. To meet the difficulty of remembering whether the figures read from left to right or *vice versa* he devised a method all his own. He noted on his left thumb a little red spot; this was the clue, dubbed by him "the knowing spot." Out of doors, as he came to a number, there would be a hasty turning back of his right or left glove to find "the knowing spot" to enable him to read the figure. Taken out by me sometimes with a sketching party he went the rounds among the young students who made much of him, plying him with raspberries. Thirty-seven years later one of these students who had not seen him since those days writes: "I can still see him with his big eyes and his sailor hat and his interest in the painting women. I think I must have sensed that he was a very unusual little boy since I remember him so well."

When he was about six years old his quite original comments on or descriptions of things led me to suspect unusual gifts and the likelihood of his becoming an inventor: hence my retention of so many school reports and letters. Though his views were so original, there was a telling simplicity and finality about them. "Nannie" who looked after him at this time, has written after an interval of some forty years: "The thing that stands out most in my mind was his *integrity* and his *intelligence* for a child so young as he then was, also you couldn't camouflage anything from him. I remember one day Alan and I playing together. I played so that he should win, but he spotted it. There was commotion for a few minutes: I must have played my part very badly . . . He used to like to see how things were done. When he was staying with us, Mr. Lee used to take him out in the mornings. He would have preferred to stay in to see how the dinner was prepared. He used to say he would be a doctor."

I wrote in 1917 to my husband: "Alan has the most delightful way of expressing himself. He announced yesterday, 'the rhubarb has made my teeth feel as if the white has come off.'" Even at five he had an answer to one's arguments which could not be gainsaid – a foretaste of his powers of argument in later years. My husband a year or so later was remonstrating with him for having the tongues of his boots all twisted and said, "Come on, those tongues should be as flat as a pancake." Quickly came back the reply, "Pancakes are generally rolled up!" As a little boy he enriched our vocabulary with new words, e.g., "quockling" for the noise made by seagulls wrangling over some booty; "greasicle" for the guttering of a candle caught in a draught. We always retained his word "squaddy" for the squat and square. Taken at about six and a half by his aunt to

see *Where the Rainbow Ends*, he was terrified when the demons
came on, so he stood up in the front of the stalls with his back to
the stage and begged his aunt, "Tell them to pull the blind down."

When it came to reading he was not very responsive to teaching.
However, finding a copy of *Reading without Tears*, he discovered
the principle (as he said later) and taught himself to read in about
three weeks. It was characteristic of his lifelong habit of achieving
things in his own way and working from first principles. In the
summer term of 1918 he began going to a pre-preparatory school,
St. Michael's, St. Leonards-on-Sea, where he does not seem to have
been an apt pupil: there were repeated complaints in his reports of
his untidiness. In later years he related with much hilarity his first
attempt there at Latin. A book of English into Latin was put before
him with instructions to look up the words in the vocabulary, and
write down the translations. Where the definite or indefinite article
occurred there was a footnote "omit." This he read as "ōmit," and
took to be the Latin for "the" and "a"; the supposed translation
was regarded by him as a kind concession. Thus his exercise ran,
"omit insula, omit mensa," etc. Despite adverse reports he made
an impression on Miss Taylor, the Headmistress. On his leaving
the school at about nine years old she said to me, "I have had
clever boys and hard-working boys, but Alan has genius." This she
deduced from the fact that he would see in a flash the solution to a
problem which other boys had to work out laboriously on paper.

Instances come to mind of instantaneous intuition. On one occasion at Sherborne in the examination for a mathematics prize, in
answer to the question, "What is the *locus* of so and so?" his answer
was, "The *locus* is such and such." This caused some consternation
since the examiners expected a page and a half of proof. Eventually
he was given the benefit of the doubt and awarded the Plumptre
Prize. When I asked why he had shown no proof he replied that
all that was asked, was, "What *is* the *locus*?" He saw it at once and
wrote it down. Eventually at Sherborne his short cuts became a
recognized feature of his work. On one occasion some examining

body sent some of his papers back to Sherborne for "interpretation," as it was not clear whether the short cuts were achieved by knowledge. The staff then had the labour of working out the intermediate steps. When he was about seventeen or eighteen his advice on the lighting of a certain hall was sought by the father of his school friend, Victor Beutell. Alan immediately gave the necessary formula. Mr. Beutell naturally asked how he knew it. All Alan could say was, "I don't know, but I will send you a proof in a few days." A great deal of this he worked out perched on top of a stepladder on the lawn in our garden. One of his Sherborne masters agrees with my surmise that Alan had a sort of "periscopic" mind which enabled him to see beyond intervening arguments to some conclusion. Changing the metaphor, my guess is that this enabled him to proceed in kangaroo-like jumps which may have been one cause of his being sometimes a "difficult author to read." On the other hand he could, as in the case of his papers on intelligent machinery, write with such lucidity as to bring the matter within the comprehension of the uninitiated.

That he took the place of intuition seriously is shown in his paper sent in 1938 to the London Mathematical Society; on pages 214–215 of this paper, "Systems of logic based on ordinals," he writes: "Mathematical reasoning may be regarded rather schematically as the exercise of a combination of two faculties, which we may call *intuition* and *ingenuity*. The activity of the intuition consists in making spontaneous judgments which are not the result of conscious trains of reasoning. These judgments are often but by no means invariably correct (leaving aside the question of what is meant by 'correct') ... The exercise of ingenuity in mathematics consists in aiding the intuition through suitable arrangements of propositions, and perhaps geometrical figures or drawings. It is intended that when these are really well arranged the validity of the intuitive steps which are required cannot seriously be doubted."

But to return to his childhood. When he was six he was taken to tea at Hazelhurst, Frant, his brother's preparatory school. Being very observant he noticed a gooseberry bush among the shrubs beside the front steps, so he asked the Headmaster if this bore any gooseberries. The Headmaster denied the existence of such a bush but Alan, being then no respecter of persons, held to his ground, and accompanied by the Head went and pointed to the bush. Like many children he would epitomise a situation in some succinct phrase. Thus, when aged seven, after chatting to my husband who was in a nursing home owing to a broken collar bone, Alan remarked, "You don't seem at all sick in your head, Daddy."

Before he was seven he had carefully concocted a mixture in which the chief ingredient was pounded dock leaves for the cure of nettle stings, the formula for which he wrote down in all seriousness with a sense of its importance. Next he began to compile an "encyclopaedio" (*sic*) for his children. Waking early he used to sit up in bed to record sundry geographical measurements, e.g., the width across England. He had a special fondness for maps and asked for an atlas as a birthday present. What little reading he did at the age of seven was of nature-study books, *Eyes and no Eyes*, etc., which perhaps prompted his remark, "there's a great deal of nature in a wood." On holiday in Ullapool, Ross-shire, in the summer of 1919, he carefully observed the flight of wild bees, which he tracked to their nests to get the honey. His head would be seen bobbing just above the deep heather as he rushed wildly over the braes, delighted to bring back some rather grubby brown honey to share with the family. When we went to a loch too distant for his short legs he would spend the day sailing his boat – a log with an improvised mast – in the little pond in the "park," as the grounds surrounding our rented lodge, Moorfield Cottage, were locally called. He seemed never tired of trying to get the *Lady of the Lee Shore* to alter course. Alan also took keen interest in watching John fishing, and did some deep sea fishing himself. Macdonald, our gillie, a delightful and shrewd companion, sized up the boys: of John he said, "Master John is a verra wise boy"; of Alan he noted with amusement, "Master Alan always thinks his fish the best and shiniest." At this time Alan derived great enjoyment from my reading of *The Pilgrim's Progress* to him and John. He was quite gripped by the tale of Christian's adventures, particularly the encounter with Giant Despair, which made him all eager for the eventual escape from Doubting Castle. On one occasion, coming to a long theological dissertation I said in an aside, "I shall just skip this"; thereupon a storm broke out. Alan, exclaiming, "You spoil the whole thing," dashed out of the room and up to his bedroom where he worked off his indignation by stamping up and down.

In the following autumn my husband and I sailed for India. Alan's letters to us between the ages of seven and nine tell of experiments with what was to hand in the garden. Thus he wrote that he had prepared a step-powder to colour steps green, but suggested no reason why one should want green steps. There were promises of a "special smoking mixture" for his father and a clay house "painted yellow as if in the moonlight" for me. These fabrications never reached us. They showed a certain amount of variation from week to week – thus "gobletoe drink" compounded of roots of grass, radish leaves and nettles and pronounced "very sweet" became two weeks later "gobletoe any grease." He also wrote out conversational captions for advertisements of Dunlop tyres for me to illustrate: as he felt confident that Dunlop would be sufficiently appreciative to send him free bicycle tyres, he thoughtfully provided the measurements of his wheels.

When eight years old he set out to write a book entitled *About a Microscope* – the shortest scientific work on record for it began and ended with the sentence, "First you must see that the lite is rite." At the end of 1920 Mrs. Ward writes of him as reading to himself very little and pottering with little bits of paper and old matchboxes, not doing anything for very long together. He was learning to bicycle and was able to ride round and round the lawn; but when someone called to him telling him to get off and come indoors, back came the reply, "I can't get off until I fall off."

When I came home in the summer of 1921 I found Alan very much changed. From having always been extremely vivacious – even mercurial – making friends with everyone, he had become unsociable and dreamy. He had greatly missed my husband and myself and his brother John, who was away at a preparatory school. I decided, therefore, after the boys and I had spent the summer holidays in Brittany, to take him away from his pre-preparatory school, where he was not learning much anyway, and teach him myself for a term and by attention and companionship get him back to his former self. Even when he was a full-blown mathematician he gratefully recalled my having made clear to him my version of the principle of long division, for as a child he always sought to know underlying principles, and apply them. Having at school learnt how to find the square root of a given number, he deduced for himself how to find the cube root.

When Alan and I were spending the autumn of 1921 in London together, rather than take him into stuffy shops, I often left him to wait outside. Regardless of traffic he occupied himself collecting with a magnet metal filings from the gutter. What he did with these I do not know, nor indeed how he came to expect metal filings there. It was, I think, in the following summer holidays he set me one of his many posers – "Mother, what makes the oxygen fit so tightly into the hydrogen to produce water?"

After Christmas holidays spent skiing at Campfer near St. Moritz he went in January, 1922, to his brother, John's, preparatory school, Hazelhurst, Frant, near Tunbridge Wells, of which Mr. W.S. Darlington was Headmaster. John was then head of the school and Alan bottom. The former was somewhat disgraced by his younger brother spending most of his spare time his first term making paper boats, possibly as a form of escapism, since boarding-school life deprived him of his usual occupations. Gym competitions at first held considerable terrors for him, but next year in his Swedish team he had the best individual score. At the end of his first term his report said he was rather pugnacious, but this complaint did not recur.

The summer holidays of 1922 we spent at Baddidorrock House, Lochinver, chiefly for the loch fishing in which Alan sometimes joined. He took a great interest in the mountains round, knew all their names and did a bit of climbing. He was keen to climb: he had heard before we went to Campfer a lecture on Everest and had wished to volunteer forthwith for the next expedition. In the long summer evenings we enjoyed dessert after dinner, picking and eating gooseberries in the garden, while we indulged in the vulgar sport of seeing who could blow the discarded gooseberry skins the farthest. Alan, applying scientific methods outwitted us all, for he inflated his discarded skins, which then far outstripped ours. I can still see how they soared over the high hedge. Here our gillie noticed that he had a natural facility in handling the tiller as the two sailed across Lochinver.

The autumn saw the boys back at school – John back to Marlborough where he had gone the previous term and Alan back to Hazelhurst; we returned to India, but he hated these partings as much as we did and we were left with the painful memory of his rushing down the school drive with arms flung wide in pursuit of our vanishing taxi. However, he soon settled down. He did not then care for outdoor games. He believed that it was at his preparatory school that he first learnt how to run fast, for he was always so anxious to get away from the ball. But he much enjoyed acting as linesman, for it afforded opportunities to calculate the angle at which the ball went out. While his good work as linesman was often praised, an end-of-term topical song contained the couplet –

"Turing's fond of the football field For geometric problems the touch lines yield."

Later he was commended for his keenness at cricket, though he does not seem to have been particularly good. He very much

Aged 5

enjoyed acting in the school play and laboriously wrote out for me the whole of one (very feeble) play. In games of hockey he was to be seen half-seated on the end of his hockey stick, and the fact that he took his games shorts to Matron for repair of the seat tells its own tale. Yet another song at the end of term poked fun at his "watching the daisies grow" during hockey. This drew from me a pen-and-ink caricature of Alan, leaning on his stick with his back to what was going on near the far goal and bent double examining a daisy in the grass. This caused much amusement at his school, not least to Alan himself. What was my surprise when Matron, Miss Dunwell, returned it to me in 1955, saying she had kept it all those years in the belief that one day a biography would be written of Alan. (I could not but be struck by the connection of that joke about "watching the daisies grow" with Alan's research towards the end of his life in the mysteries of growth.) Encouraged, however, by accidentally hitting a goal he was soon playing quite well.

At Hazelhurst he took up chess and stirred up some interest in the game among his fellows. Having settled down happily at school he seems to have been extremely talkative. One way and another he provided material for the end of term topical song, e.g., with his homemade fountain pen which "leaked enough for four." His letters from 1923 onwards describe with very clumsy diagrams his inventions, typewriter, camera, etc., and he once wrote with a specimen of his patent ink. Mrs. Darlington, wife of his preparatory school headmaster, writing in May, 1923 remarked that his inventive brain was always busy. From the "inventions" of his earlier days it can be seen that his brain was occupied constructively, so it is little wonder that he was often on the library "black list" for not borrowing books of fiction. Records of the Debating Society contain some typical remarks. Opposing the motion, "It is more interesting to keep pets than cultivate a garden," one boy said that flowers could be taught to climb; Alan objected: "once they climb up they can't climb down." Another proposition was, "Electricity is more useful than gas." Alan opposing, characteristically said,

"Air is gas and so necessary for life." About a year before Alan left Hazelhurst Mr. Darlington wrote to me, "John is of the vintage that appeals to anyone of discernment. Alan is the caviare; but to one who knows him as well as I do the appeal is as strong as John's."

After our return to India at the end of 1922, John and Alan began to spend their holidays with the Rev. Rollo Meyer and his wife at Watton-at-Stone, Hertfordshire. They fitted very happily into the family life there and Alan enjoyed bicycling round the country and carrying out various experiments in a neighbouring wood. Letters tell of his coming in black all over and of his singeing his eyelashes as he fired the clay pipe he had made. Mrs. Meyer wrote later that he was always doing dangerous things. It was at the end of 1922 that his eyes were really opened to the world of science by the gift of *Natural Wonders every Child should Know*, by E.T. Brewster, a book well worn and greatly valued by him and of which even in his last years he spoke highly.

With John he spent part of the summer holidays of 1923 with M. and Madame Godier in Rouen. Though he found the language difficult he persisted in talking and seems to have made his way all right, for Mme. Godier describes him as, *"Alan, si charmant."* This visit started his interest in French and he plunged forthwith into writing to us in French – of a sort. He was becoming less detached at school. This detachment I believe to have been due to pre-occupation with his "inventions." At his preparatory school from time to time he was mentioned for good work and managed to keep towards the top of his forms. His conduct, scripture, French and mathematics were well reported on, but his work was often marred by bad writing and untidiness. His writing was so appalling that in the Easter holidays of 1924 when he was nearly twelve he and I settled down together to reform it, and for a time he took great pains and improved his writing beyond recognition, but by the end of the year it was reported to be "as bad as ever."

As a small boy he showed the certainty that he later displayed; where others would say, "I always thought such and such," he would say, "I always knew." He maintained that he "always knew" that the forbidden fruit in the Garden of Eden was a plum. He experienced quite unaffected concern when his mathematics master in an endeavour to initiate the other small boys in his form into the mysteries of algebra had pinned "x" down to something much too determinate and concrete for Alan's dawning logician's mind.

We spent the first part of the summer holidays of 1924 in Oxford. Here he was in clover having been given a set of chemicals with instructions for various experiments, on which he spent many happy hours in the summerhouse. For exercise he enjoyed riding lessons and exploring the countryside on his bicycle. We passed the latter part of these holidays up at Festiniog, in North Wales, where his interest in climbing grew. Together he and I climbed neighbouring peaks and place Snowdon. We had some idea of going to Knocke in Belgium. If we did he thought he would like to cycle into Germany to climb a mountain higher than Moelwyn. After our climbs he busied himself next term making a map of the places we had gone to (with names we had made up for them) and we were asked to compare his map with the ordnance map. He took a special interest in the ordnance maps we used, having learnt something from his father about surveying, and having read about how heights of mountains and contours were calculated. For a time map-making became one of his hobbies. In 1924 my husband resigned from the Indian Civil Service and we spent the winter and spring of the following three years at Dinard in France.

In November, 1924, Alan wrote, "I have come into great luck: there is an encyclopaedia that is first form property." For his Christmas present, 1924, we set him up with crucibles, retorts, chemicals, etc., purchased from a French chemist. Hours were spent in the cellar: asked one day where Alan was I replied, "At home in the cellar experimenting. I am sure he will blow up himself and the house one day." Now, at twelve and a half, serious research began. Quantities of seaweed were brought back from the beach for his experiments in making iodine. Seeing his eagerness we asked a Mr. Rolf,[1] a schoolmaster from Shrewsbury, to give Alan some coaching in science. Evidently he was much impressed by the boy's aptitude and knowledge and told friends of ours about this boy who seemed to have been born knowing the properties of elements.

Alan and I often talked about great scientists, particularly Pasteur whose portrait was on the French stamps, and I set before him the possibility of himself making some great contribution to science. With the help of the encyclopaedia he was trying to learn organic chemistry by himself (at twelve and a half), and one of his letters contains graphic formulae showing molecular arrangement in alcohol, methyl ether, etc. In March, 1925, he was occupied in preparing questions for Mr. Rolf whom, unfortunately, we did not meet again. At this time in a letter he says, "I always seem to want to make things from the thing that is commonest in nature and with the least waste of energy," and he mentions making a list of experiments in the order he means to do them.

Alan had an extraordinary gift for winning the affection of maids and landladies on our various travels. Our old Breton cook-renowned for her *"sale caractére"* – to use her own words – was so devoted to him that her jealousy of me became quite a nuisance. When Alan used to come up from the cellar where he had been experimenting she would press him to wash his hands in the scullery to which no one else was allowed entry. Holidays in Dinard, where he had some lessons in French, spurred him to writing his home letters once more in French, with unexpected repercussions on one occasion. Quite unaware that a post-card might be read by anyone other than the addressee, he wrote *"Vous souvenez de la revolution qui peutêtre se trouverait cette terme. Je pense que M. Darlington a pensé comme nous et cru qu'il fasse qu'il répande les offendeurs."* (The *revolution* was no more than some uppishness among a few boys.) Mr. Darlington read the card and naturally asked for an

---

[1] Spelling doubtful.

explanation. The letters in French teemed with mistakes and odd circumlocutions but quite clearly conveyed his meaning; indeed he wrote more fully in French than in English. He asked for cuttings of interest from French newspapers. I pressed hard for Alan to try for a scholarship at a public school, but his father and headmaster objected supposing he would disgrace himself and his school. Alan nearly always shone in examinations, as was demonstrated when, in the autumn of 1925, trying for a chance vacancy he took the Common Entrance Examination for Marlborough College in order to join his brother John; Marlborough College would have placed him in the "Shell" had there been a vacancy and according to Mr. Darlington he quite surpassed himself. In the Lent term he again took Common Entrance, this time for entry into Sherborne School the following term, and was placed in the Shell.

3

At Sherborne School

The General Strike of 1926 broke out as Alan's first term started. He dearly loved a bit of adventure and so was delighted with the opportunity the strike gave of a novel mode of arrival at school – a new school at that. Landing at mid-day at Southampton from France, he sent a telegram to his housemaster, Mr. Geoffrey O'Hanlon, promising to report next day. Then he disposed of his trunk, and set out armed with a map to bicycle to the school. When he hopefully suggested the possibility of bicycling, I insisted that he should not attempt the whole sixty miles in one day, thinking he might be ruthlessly made to attend early school next day. He spent the night at Blandford at the Crown Hotel, where he seems to have caused some diversion, for the whole staff turned out next morning to see him on his way. His bill for dinner, bed and breakfast was purely nominal – six shillings. This unusual way of arrival won him some notoriety and was even reported in the local press. His enterprise stood him in good stead a year or two later; when progress was at a low ebb his housemaster, seeking some redeeming feature remarked: "Well, after all he *did* bicycle here."

Alan's first letter from Westcott House, Sherborne, informs us that "Mr. O'Hanlon is very nice" and indeed he proved to be the perfect housemaster, to whom we owe an undying debt of gratitude. Years later in a circular letter to present and Old Boys, announcing his intention of giving up Westcott House, he disclosed the aim he so notably achieved. "*Humanitas,* if you recollect any Latin, combined with a sense of humour, represents most of what at the back of my mind I wanted to put before you. I don't think you will accuse me of attempting to force a pattern on you: the attempt from me would have been conceited and ridiculous." It was, surely, this profound respect for the individual, be he but a schoolboy, which enabled him to judge how far to tolerate, and how much to prod and exhort, the boy whom after a week he described as *sui generis*.

Alan took an early opportunity to exhibit his precious iodine made from seaweed to his science master, somewhat to the latter's amusement. Alan's whimsical humour comes out in his second letter from Sherborne, when he compares the fagging system to that of the "Gallic councils that tortured or killed the last man to arrive; here one fagmaster calls and all his fags run, the last to arrive getting the job." (It was a touching tribute that Mr. Arthur Harris, to whom he had been fag, took the trouble to come from London to Woking for Alan's funeral.) In the "Shell" he was working with boys a year older and holding his own, even in Greek, a subject new to him. In general the two reports of his first term indicate good work. Mr. G.B. Sleigh stated: "He is delightfully ingenuous and 'unspoilt': he shows distinct promise. His Latin is good for his age in spite of some inaccuracy," and his housemaster summed him up as "original: a very interesting character." He won the Kirby mathematics prize for the Lower School in his first term.

The summer holidays of 1926 were spent by Caragh Lake in County Kerry. My husband and John fished assiduously, Alan did so only in a desultory fashion, but nevertheless caught the largest trout of the season – $15\frac{1}{2}$ inches long. He and I climbed Carntual, the highest mountain in Eire, and also one of its spurs, Benkeragh, preceding the latter climb with a bicycle ride of some few miles.

His housemaster's report at Easter 1927 runs: "His ways sometimes tempt persecution: though I don't think he is unhappy. Undeniably he is not a 'normal' boy: not the worse for that, but probably less happy." After his promising start there was a sad falling off. He seemed to find Greek difficult and his French showed fluctuations from good work to inattention, "except," characteristically, "when something amuses him." Mathematics, in which he started very well and promisingly, by the beginning of the summer term of 1927 were "not very good. He spends a great deal of time in investigations in advanced mathematics to the neglect of his elementary work." In science "his knowledge though scrappy" was said to be considerable. At the summer half term, 1927, his housemaster discerningly complains of his "trying to build a roof before he has laid the foundations," and diagnosed that his head was a little turned by his science. The deterioration in work was the more deprecated because, at my husband's request, he had been generously allowed to substitute golf for cricket. It was possibly not very good for him, for I can imagine him thinking out problems rather than minding much how he played.

At half-term he got mumps in a kind of suppressed form which seemed to sap his vitality, and he only emerged from the sanatorium just about in time for some terminal examinations.

Nevertheless he carried off the Plumptre Prize for mathematics and did a remarkable examination paper which won from Mr. (now Colonel) Randolph the tribute, "A mathematician I think." As Alan was nearing fifteen years of age he discovered Gregory's Series for "$\tan^{-1} x$," unaware that it was already well known. It was a cause of satisfaction to Alan himself that this was achieved by him without any knowledge of calculus. On his asking if the series was correct, Colonel Randolph, his mathematics master, at first thought Alan must have got it from a book in the library. This master clearly recalls the incident and writing thirty years later he adds: "His form-master complained that his work was so ill-presented that he ought to be sent down. Mr. Nowell Smith asked me for my opinion and I replied, 'The boy is a genius and ought to be sent up.' With typical British compromise, he was of course left where he was. I still don't know whether he could not or would not produce on paper the arguments by which he got his answers. I remember so well asking how he got the answer to a rather difficult problem. He could only say, 'Well, it's right, isn't it?' and of course it was."

Once more in the summer of 1927 we were back in Festiniog, in Wales, to fish and climb. A guest in the same house was a certain Mr. Neild who took a great interest in Alan and gave him a book on mountain climbing, in which he wrote a long inscription treating Alan's climbs as symbolic of his eventually attaining intellectual heights. Looking back I realize that Alan must have been pretty sturdy for a boy of just fifteen, as we climbed several peaks in Snowdonia – Tryfan, Glyder Fawr, Carnedd Llewelyn as well as Moel Siabod and Cader Idris. Most of these expeditions involved a sequence of bicycle rides, train or bus journeys followed by a walk before beginning the ascent: all this against time to fit in with returning trains or buses. We again climbed Snowdon, but were caught in dense mist and had to spend the night on hard chairs in the restaurant, the hotel not being open. On the way up Moelwyn we met Wilfred Noyce, then a little boy, striding up a shoulder of the mountain ahead of his father and brother. Alan's comment later when Mr. Noyce became a famous mountaineer, was that this was the only occasion on which we had climbed by a route more difficult than that followed by Mr. Noyce.

Next term he was described by another master as "a keen and able mathematician," and, "a very interesting boy to take: he is full of ideas and keenness." After the admonitions of the previous term he began to show some improvement, but untidiness was still emphasised. Mr. Hornsby Wright's criticism of his science was, "his methods in practical work are at present deplorable." He still relied on a special spurt in examinations to pull him through. Mr. O'Hanlon seems to have regarded with rather amused detachment Alan's way of earning a bad half-term report and then coming out top in the terminal examinations. Thus, seeing Alan on one occasion when this had happened he called down the corridor to him, "Well, Turing, so you've done 'em down again," and his report runs: "He did very good exams. and confounded many folk ... he certainly has ideas and imagination." He also saw signs of improvement in personal appearance. At the end of the Michaelmas term 1927 he says, "No doubt he is very aggravating: and he should know by now that I don't care to find him boiling heaven knows what witches' brew by the aid of two guttering candles on a naked windowsill. However he has behaved very cheerfully; and undoubtedly has taken more trouble, e.g., with physical training. I am far from hopeless." As regards "the witches' brew" Alan's only regret was that Mr. O'Hanlon had missed seeing at their height the very fine colours produced by the ignition of the vapour given off by super-heated candle grease. This was not the only instance of his taking his housemaster by surprise. On return from Chapel one Sunday, Mr. O'Hanlon found a weighted string suspended from the staircase wall. This had been fixed up and set swinging by Alan before Chapel to demonstrate the rotation of the earth by the change of direction of the string during the forenoon. Though the experiment was primarily in the interests of science, there may have been an impish desire to test how far he could go, without breach of discipline.

On 7th November, 1927, he was confirmed by the Lord Bishop of Salisbury and Mr. Nowell Smith expresses the hope that taking this seriously he may not be content to neglect obvious duties in order to indulge his own tastes however good in themselves. Later he wrote, "He is the kind of boy who is bound to be rather a problem in any school or community, being in some respects definitely antisocial. But I think in our community he has a good chance of developing his special gifts and at the same time learning some of the art of living." And here perhaps I may pay my tribute to this great headmaster's patience, discernment and liberal attitude towards the boy whom he had styled "The Alchemist." Not many headmasters would have tolerated Alan's failure to fit at first into public school life. That he seemed for a time anti-social (though I would say rather non-social) was, I imagine, more due to shyness or abstraction and absorption in his own cogitations than to any spirit of antagonism. He may have been unconsciously a passive resister.

Mr. Nowell Smith resigned at the end of 1927. On his retirement he wrote to me describing Alan as "out of the common run" and as a boy of "marked originality." Before he left, Mrs. Nowell Smith, who knew many boys outside the Headmaster's House, wrote thus to me: "We shall follow your boy's career with very great interest. He will do something great in science I am convinced. Every time I have come across him, even when he helped me to weed in the garden, I had a feeling of his power. I daresay he is very often aggravating, I don't know that he is . . . but particularly with great scientific men they have a childhood like your boy's." Before Alan went to Sherborne I had met Mrs. Nowell Smith and given some hints about what to expect. She contrasted my description with the more favourable accounts given by other parents of their boys. Though he had been loved and understood in the narrower homely circle of his preparatory school, it was because I foresaw the possible difficulties for the staff and himself at a public school that I was at such pains to find the right one for him, lest if he failed in adaptation to public school life he might become a mere intellectual crank. It will be seen later how Sherborne School justified Mr. Nowell Smith's hopes and mine.

In the Christmas holidays of 1927, Alan, being then fifteen and a half wrote for me a précis of one of Einstein's books on Relativity in order to help me to understand the subject. I still have this rather untidily written précis. It shows him surprisingly conversant with the work of other scientists and mathematicians, doubtless due to those excursions into advanced mathematics which, as his masters said, were carried on to the detriment of elementary ground work. Even at that age his independence of thought is illustrated by the conclusion in his précis of one chapter as follows: "You will notice these results are those of Chapter XII. I have gone rather differently from the book because in this way I think it should seem less 'magicky.'" He does his best to "temper the wind to the shorn lamb" with parentheses to make it easier for me. Thus a note to Chapter VI runs, "The words' relative, relatively to,' etc., are not used in the sense of proportion. Perhaps the words 'as measured by the system of' substituted for 'relatively to' will make it clearer what is meant." Encouragement is given by appending to his précis of Chapters VIII and IX his view that "the arrangement is fairly simple." Considering one of Einstein's proofs complex Alan writes, "I have got a proof though which will give you the result of Chapter XVII directly and if you like the Lorentz transformation." Further on he adds, "Now read Chapter XVII. It should interest you particularly." I am recommended just to read Chapter XV through (as it does not need very careful reading) and to think about what interests me. Here are more quotations: Chapter XX. "The explanation that a gravitational field is stopping the earth from moving past the train seems very arbitrary, but then people do not as a rule use railway carriages as their systems." Chapter XXIX. "He (Einstein) has now got to find the general law of motion for bodies. It will have of course to satisfy the General Principle of Relativity. He does not actually give the law which I think is a pity, so I will. It is: 'The separation between any two events in the history of a particle shall be a maximum or minimum when measured along its world line.'" I am sure it never occurred to him that there was anything unusual in a boy of fifteen and a half elucidating Einstein on Relativity, for while all this was simmering in his brain he was very jolly and lively in the family circle.

The school reports remind me of the incoming tide – showing a sort of rhythm of progress and recession. The recessions were what chiefly impressed my husband, who made a rule that Alan's reports were not to be opened at breakfast. Not until my husband had been fortified by *The Times* and a couple of pipes could he face them. Alan's comments were, "Daddy should see some of the other boys' reports," and "Daddy expects school reports to be like after-dinner speeches." It would have made my husband open his eyes could he have foreseen the day, some twenty odd years later, when a future Headmaster of Sherborne School would be invited to dinner to meet Alan.

Aged 5

In 1928 the time came round for taking the School Leaving Certificate. There was considerable tension in the common room between the literary and the scientific members of staff; the former maintained that Alan was quite unqualified for the examination, the latter protested that he should not be held back. The Headmaster, Mr. C.L.F. Boughey, gave the casting vote to permit him to have a try at a term's notice. Coaching by his father in the Easter holidays in English and by me in Divinity and some private tuition in Latin in term time were given to help him on. But Mr. Bensly, who had a special form – called by him the "Vermisorium" – for School Leaving Certificate candidates, promised to give a billion pounds[1] to any charity named by Alan should he so much as pass in Latin. However the usual spurt was made and results showed credits in seven subjects, including Latin, English and French, much to the delight of Mr. O'Hanlon who added to his congratulations some digs at Alan's detractors. And now, after this success, things began to look up somewhat in that he was sociable and more friendly. It was probably about now that he became "Mathematician-in-ordinary" to his house and was in demand for help with other boys' prep.: this possibly led to more interest in his fellows and making friends. His help to other boys was probably why Mr. O'Hanlon speaks of his being unselfish in temper. In the upper school he was able to concentrate on mathematics and science. Mr. Eperson[2] reports of the former: "He thinks very rapidly and is apt to be 'brilliant' but unsound in some of his work. He is seldom defeated by a problem, but his methods are often crude, cumbersome and untidy." English is still under criticism for weakness in reading. He could be neat when not too impatient to get on and pursue his train of thought. For instance, at about sixteen he filled an old-fashioned spherical glass gas lampshade with plaster of Paris, covered it with paper and neatly marked the position of the principal stars and

[1] Note added for this edition: worth many times more than it is today!
[2] Now Canon Eperson.

constellations from his own observations. This had an amusing upshot: once, suspecting a burglar at 3 a.m. I ventured downstairs only to find Alan about to go out to stargaze and complete work on his globe. His originality was applied to drawing. In one of the school art exhibitions was a picture by him of the Abbey showing the upright lines sloping in perspective: Mr. Cecil Hunt, the artist invited to criticize the exhibition, singled out this drawing as being of particular interest.

It was probably at about sixteen or seventeen that Alan indulged in sundry escapades for wagers: thus in mid-January he swam the Yeo and in June went on O.T.C. parade in overcoat and no tunic. It was only long afterwards that we learnt of these pranks from an old Shirburnian who met John abroad in the war. There were rumours of some traffic, along with a partner, in ice creams, until quashed by his housemaster. These unorthodox adventures point to a light-hearted attitude to school apart from study. Before he left Sherborne he was running for his house in steeplechases and playing football really well.

His housemaster assures me of the sympathetic attitude of many of the staff and of the general friendliness and tolerance shown by his contemporaries in the house, who appreciated him as a member of the community and as a friend. Among these several names stand out in my own memory-those of Peter Hogg, George Maclure, Blamey, Addison, and Victor Beutell. He kept up after schooldays with some of these and in 1933 went on a walking tour with Peter Hogg and George Maclure. They put up in all sorts of strange places along with tramps. Fairly soon after they went up to their respective universities Peter Hogg, who had shared a study with Alan at Westcott House, wrote from New College: "Thanks for your most heartening epistle. I love you when you get excited: your spirit flashes up from beneath the cinereous growth of dirty untidiness that you usually hide it with."

A glimpse has come of the impression made on one of Alan's younger schoolfellows: he was in the dormitory in which Alan, as head, enjoyed the privilege of a candle to read by after "lights out." This Old Boy wrote to Mr. O'Hanlon saying that once he mischievously blew out Alan's candle and did not get the beating he deserved. It was pointed out to him by Mr. O'Hanlon that he had the distinction of doing this to a genius classed with Einstein. Whereupon the reply came back, "I was always a bit in awe of his 'brain,' but chiefly remember him for his good humour and gentleness. Surely we knew that he was a genius even then: I know I did when I was told he was teaching Ben Davis in some high mathematical way: and I remember a globe in his study showing the constellations." As Mr. Davis was the Senior Mathematics master who took the VIth Form and Army Class mathematics Alan can have been only showing him something of his mathematical findings, hardly teaching.

Alan kept a private locked diary. Unfortunately one boy out of mischief or from some other motive forced the lock and, I believe, irreparably damaged the book, in which was probably entered mathematical research. This piece of wantonness has deprived us of valuable records from which his early development might have been traced. The loss very much distressed Alan.

In the summer of 1929 Alan passed the Higher Certificate despite the fact that he forfeited the marks for his best answer because he had answered one question too many. Next year he also passed, this time absent-mindedly removing with his rough paper one of his answers. When he passed yet again next year he won distinction, though rather handicapped by having German measles during the examination. In December, 1929 he took the scholarship examination at Cambridge, but as he only qualified for an exhibition my husband decided he should remain another year at school and try again. He occupied his leisure time at Cambridge making a plan of the town. Shortly before the examination he had written to me about Schrödinger's quantum theory, but whether this was within the scope of his normal work or the result of his probing into advanced science I do not know.

With some hesitation Mr. O'Hanlon in 1930 promoted him to be a house-prefect and in a letter to me wrote: "That he will be loyal I am well assured: and he has brains: also a sense of humour. These should carry him through... His fag is not for his own dressing or the assortments or ill-assortments of his study. Half his duty I have come to the conclusion is being able to say 'no' to himself and others at the right times." This responsibility brought him out, but his absentmindedness often caused consternation to fags whom he despatched on errands forgetting to provide the necessary money. The extra time at school with a certain position helped to develop him on the social side so that Mr. O'Hanlon at the end of 1930 could report, "A very good term. He wins respect both by brains and character," while the Headmaster (Mr. C.L.F. Boughey) confirms this with the comment, "He is a distinguished and useful member of the community." As things turned out Mr. O'Hanlon came to realize that his confidence in Alan was fully justified. He found that Alan took a fatherly interest in the boys in his dormitory and "no doubt imparted his learning and curiosity to them."

In the close friendship and memory of his friend, Christopher Morcom, Alan found special inspiration, but this boy who had just won a Natural Science Scholarship to Trinity College (Cambridge) died in February 1930. How deep was his attachment is revealed to me in a letter at the time. "I feel that I shall meet Morcom again somewhere and that there will be some work for us to do together as I believed there was for us to do here. Now that I am left to do it alone I must not let him down, but put as much energy into it, if not as much interest, as if he were still here. If I succeed I shall be more fit to enjoy his company than I am now." Mature thoughts for one so young. He continued, "It never seems to have occurred to me to make other friends besides Morcom, he made everyone seem so ordinary." How poignant and deep was his grief may be gathered from the fact that all his life he kept his father's and my letters to him about his friend's death as well as a large number of letters from Morcom, mostly about science and especially astronomy, and also many letters from Mrs. Morcom and notes of all that had been said about Morcom. Mrs. Morcom wrote that Alan's letter to her had helped her more than anyone's. He did not however let his grief get the better of him. His terminal report was particularly good – the best yet – and his housemaster was especially appreciative. He put his back into everything – work and sport – and came in first in the house steeplechase. In the Easter holidays following this sad term he went with Colonel and Mrs. Morcom on a trip to Gibraltar and Granada, and with Mr. O'Hanlon's party of boys to Rock in Cornwall, leaving only a short spell at home. When in Cornwall he daringly jumped across the precipitous Tregudda Gorge, Padstow, and almost slithered into the water beneath.

The Christopher Morcom Prize for Natural Science, awarded by Morcom's parents, was won in 1930 and 1931 by Alan. In connection with this award in 1930 special mention is made of Alan's original work in a paper on "The reaction of Sulphites and Iodates in acid solution" – a paper pronounced by the hard-to-please H.M.I. to be "a very creditable piece of work." Mr. A.J.P. Andrews, one of the science masters, wrote with respect to it twenty-four years after Alan had left Sherborne: "I first realized what an unusual brain Alan had when he presented me with a paper on the reaction between iodic acid and sulphur dioxide. I had used the experiment as a 'pretty' demonstration – but he had worked out the mathematics of it in a way that astonished me . . . I have always thought Alan and his friend Christopher Morcom were the two most brilliant boys I have ever taught." By the summer of 1930 his work in mathematics and science was sounder, more mature and in better style. Even in English he earned approval for producing "sound work with imagination and interest." The summer holidays of 1930 we spent at Gartan, County Donegal. Here Alan fished with his father and brother and with me climbed the adjacent mountains.

In December 1930 he went up again to Cambridge to take the scholarship examination. His brother John and I met him at Waterloo on his way through London. The examination in no degree weighed on him – for he revelled in examinations. He was exuberant and full of zest and out to enjoy himself and determined to go and see the film *Hell's Angels* on his way through. On his winning, in December 1930, an open major scholarship in mathematics to King's College, Cambridge, he was made a school prefect, an appointment of which, apparently, he proved himself worthy. One master wrote that if ever a boy got a scholarship by his own efforts Alan did. When he returned for the Christmas holidays with a great pile of prizes in his arms, to my enquiry what they were, his reply was, "Oh! I don't know: but I think when you've been a couple of years in the sixth they start pensioning you off."

On two summer holidays Alan was one of Mr. O'Hanlon's party of boys and other friends in Sark, Channel Isles, where he did some sketching in water colours and enjoyed mixed bathing at midnight. Mr. O'Hanlon describes him as a "very lively companion" on these holidays, for though shy with strangers he could be very hilarious at home and with those whom he knew well. On one of these visits he took a box of test tubes containing fruit flies which he was interbreeding: these flies all had their own names – one being dignified as "Humphrey." When his suitcase was unpacked the process of interbreeding had much advanced and I gather the experiments got rather out of hand. He had a particular aversion to the jellyfish which abounded around Sark. These, to the accompaniment of loud cries, he used to bombard with stones before bathing.

And so his somewhat varied career at Sherborne came to an end with the award of a school leaving scholarship and the Westcott House Goodman scholarship and the King Edward VI's gold medal for mathematics. It says much for Sherborne and Alan's own efforts that, after being at first something of a misfit in public school life, his housemaster could write of his bringing his career there "to a very successful conclusion," adding, "I am grateful to him for his essentially loyal help," and in a later letter he thanked him not only for this, but for the enjoyment he gave by being there, guaranteeing that Turing would be a household word until the present School generation had disappeared. The Headmaster's final report runs, "a gifted and distinguished boy . . . I have found him pleasant and friendly," and in his speech on Prize Day said, "Turing in his sphere is one of the most distinguished boys the school has had in recent years." As one of his prizes he chose *Mathematische Grundlagen der Quantenmechanik* and soon wrote of his interest in it. Alan himself held that the reason he got on so well with Mr. O'Hanlon was that his housemaster never had to teach him. More's the pity! He might have instilled neatness and produced good writing. Mr. O'Hanlon's letter to me on Alan's leaving Sherborne runs thus: "It's a sad business, writing the last of these epistles. But there's nothing sad to look back on. Alan has done really well. A couple of years ago one doubted his capacity to be a prefect: but there's no doubt of his success and he has carried it off. Mathematicians and Scientists one is apt to regard as being soulless creatures: but Alan is not, he is warm-hearted and has a saving humour. We shall miss him, for he was a character and won respect."

Perhaps the lasting impression left of his schooldays is summed up in a letter long after (in 1954) to me from Mr. O'Hanlon: "It's not however as a great brain that I shall remember him, though I acknowledge it with awe and feel a degree of humility that Westcott House should have seen him through some of his days, not without happiness. But the picture I keep and treasure is of a somewhat untidy boy arriving during the Railway Strike, after making his way on a bicycle from Southampton via the best hotel in Blandford, and reporting, 'I am Turing.' It was a good start. He was odd in various ways – as you know: as e.g., in putting butter on his porridge. But it never made him peculiar or unsocial.[3] Neither his contemporaries nor, I fancy, his masters knew he was of a calibre that a school is lucky to number in 100 or 200 years. But they liked him as a person and a character. I look back on holidays in Cornwall and Sark

---

[3] Previous signs of seeming anti-social were evidently quite temporary.

among the great enjoyments of my life: in all his companionship and whimsical humour, and the diffident shake of the head and rather high pitched voice as he propounded some question or objection or revealed that he had proved Euclid's postulates or was studying decadent flies – you never knew what was coming . . . what it was to be and remain human and lovable."

At home we regarded Alan as the family encyclopaedia; he seemed to have the answers to all our scientific queries. He was always most amusing; the most ordinary acts, such as going up to town, somehow, in his case, were fraught with adventure or unusual occurrences, so that he came in for a good deal of chaff, from me especially, but it was always taken in good part. For all his brilliance and his achievements he never posed as "high-brow."

4

At Cambridge

Being a scholar born, Alan found himself thoroughly in his element when he went up to King's as a mathematical scholar in October 1931. The balance of freedom and discipline just suited him. He at once took up rowing, being in the College trial eights in 1931, 1933 and 1934. For one so shy it was a curious thing that he very much enjoyed reading the lessons in King's College Chapel, which as a scholar he was occasionally called upon to do. He was too reticent about his religious beliefs to reveal just where he stood. He often accompanied me to church at festivals, as well as attending chapel at King's – things he was too honest to do had he not been, at least, in limited agreement with Christianity, though he was certainly not orthodox. "Within the framework of his science he believed in the great order of things," according to a very intimate friend. In his last year or so, happening to glance through the Church of England Catechism, by then forgotten by him, he expressed great admiration for the sound exposition of the "Duty to one's Neighbour." Whenever he referred to Jesus Christ he always did so with reverence using the term "Our Lord," which speaks for itself. While still at school his opinion of daily school chapel was that the practice was a good one, even if one were apt to be in a state of semi-coma. This may sound paradoxical, but what he commended was evidently the corporate recognition of religious obligation.

Soon after arrival at King's he wrote, "I pleased one of my lecturers rather the other day by producing a theorem which he found had previously been proved by one Sierpinski, using a rather difficult

method. My proof is quite simple, so Sierpinski is scored off."
This rather points to his continuing his investigations on his own
account.

On a visit to Cambridge I had breakfast a few times in his
rooms. His cookery methods caused me some concern. A wisp of
blue tissue paper projecting several inches was wrapped round the
frying-pan handle to keep his hands clean and as this came very
close to the gas flame there was the risk of setting all ablaze.

Great was my surprise when he returned one day from London
with a second-hand violin. He taught himself (except for a very few
lessons now and then): indeed he never got beyond a very elementary stage but nevertheless he derived a great deal of enjoyment
from playing it; and though he was not really musical he kept up
intermittent practice for some twenty years.

His obtaining only a second in the first part of the mathematical
Tripos (1932) suggests that he was still negligent of those immediate
tasks which did not hold his interest; instead, true to type, he may
have been indulging in more advanced investigations, while the
distractions and interests of his new surroundings may well have
contributed to this disappointing result. Following the first part of
the Tripos he went with his father on a walking tour in the Black
Forest – rather marred, however, by Alan being prostrated with hay
fever on the journey. On return he was back again as usual at King's
for the long vacation term and then went on a camping holiday
with John in Connemara. We had great trouble at the time getting
a reply from him about his movements. Eventually he turned up
at Cork having crossed in a pig-boat.

Together with several other Kingsmen he joined the anti-war
movement, and was interested in its demonstrations, but as there
is no mention of it later, it was probably a transient phase. There
is not much in the way of records relating to the period between
the first and second parts of the Tripos, that is, 1932 to 1934: life
went on much as usual with work, rowing, making friends and a
general extension of his outlook. In the autumn of 1933 he read

a paper "Something by way of Mathematical Philosophy" to the Moral Science Club.

Alan's friend, Mr. Denis Williams, however, throws some light on these undergraduate days as follows:

> Alan Turing was a year or so my senior at Cambridge and I think it was as members of the Boat Club that we first made each other's acquaintance. Rowing is traditionally the sport of 'toughs,' but at that time the King's Boat Club was an odd mixture of those born tough and of intellectuals in search of a counter-irritant. Alan's close friends were among the intellectuals, but I believe that a sport that demands the maximum of exertion held a natural attraction for him and there was a sterling quality about him which gained him respect in any company. There is a saying that 'He who drinks wine, thinks wine: he who drinks beer, thinks beer.' Alan could and did drink both, but in either event he remained entirely himself. As a beer drinker, in the literal sense, I remember he once achieved a certain distinction by swallowing a pint in one draught. This, he maintained, could not be done with water because the attempt made one disgusted with oneself. In the 1930s Germany was a popular place to go to in the vacations because of the favourable terms for students. It was my good fortune to be in Alan's company on two such excursions, once on a ski-ing party and once on a cycling tour. English students in Germany would often use the 'Heil Hitler' greeting either jokingly or as no more than the equivalent of 'how-do-you-do?' It was typical of Alan that he did not do so. So far as I remember he had no particular political affiliations, but he recognized Nazi doctrine as an evil and was not prepared to compromise with it by the smallest gesture. On one occasion the warden of a youth hostel disclosed his lack of Nazi sympathies to him; a remarkable confidence to a complete stranger. Another incident that I recall illustrates his keen nose for any sort of insincerity . . . There was nothing boorish about him, but he had no use for any sort of pretence.

Another contemporary was J.W.S. Pringle, now Dr. Pringle, F.R.S., of Peterhouse, Cambridge. He says:

> I remember him mainly for the enjoyable evenings we spent playing bridge together, and for his quality, rare among mathematicians, of being

able to help non-mathematicians with some of their problems. He had a way of being able to put into simple language difficult mathematical ideas and of seeing at once when a mathematical approach would or would not help with a particular problem. He was, I think, the first of our year to be elected to a Fellowship, and we all recognized that this was a due reward for outstanding promise.

On occasion Alan seems to have impressed his lecturers. As an undergraduate he had attended a course of lectures on Electricity and Magnetism given by Dr. Sydney Goldstein, F.R.S. Though some twenty-five years or so must have elapsed since then, Dr. Goldstein has written to me in 1957 that he still remembers outstanding work handed in by Alan. He adds, "Alan was one of a very small number – some three in all – whose undergraduate work I particularly remember in that way during the whole time I was lecturing at Cambridge." That he should have written from Haifa twenty years after his early contact with Alan to congratulate him on his Royal Society Fellowship testifies to the impression made.

Whether because the result of Part I stimulated him to more concentration on the main business or because the work for Part II was more in his line (maybe for both reasons), he showed his true form in his finals. For Part II of the Tripos he came under the old regulations which meant that in addition to Part II he took Schedule B. This time the results were all that we could wish for, he came out a Wrangler with a B star (distinction) and was awarded by King's College the Harold Fry Studentship. Professor Littlewood, F.R.S., recalls that Alan in preparation for his finals consulted him on the subject of Elliptic Functions. Though the details of their conversation are forgotten, he does "remember recording from that moment that Alan was a marked man." Between the two parts of the examination Alan dashed home to see his father, who was about to have a major operation, but had to hasten back to Cambridge to row in King's Second Boat before taking Schedule B. His boat met with no success for it ended up on the bank, as well as I can remember.

After spending part of the long vacation, 1934, at home and being best man to his brother on his marriage to Joan Humphreys, Alan was back again at King's, and now installed on the top floor (X staircase) of Bodley's Buildings, in rooms which overlooked the backs. He was busy on research: his first publication, "Equivalence of Left and Right almost Periodicity," described by Professor Philip Hall as a "very pretty little proof," was received by the London Mathematical Society in April 1935, and appeared in their journal, Vol. 10, 1935.

He did some coaching, but found that it required too much time, since he was fully occupied with writing his thesis on the "Gaussian Error Function," which won him his election into a Fellowship at King's in March 1935 – before he was twenty-three. That meant a half-holiday for his old school.

He gave a very hilarious account of his first evening as a Fellow, dining at the High Table, and afterwards of the time in the Common Room, when he felt somewhat dazed after having had more wine than he was accustomed to. However he kept his head sufficiently to win a few shillings at Rummy from the Provost. I remember so vividly his rubbing his hands together as he described the evening and boasted gleefully, "I do fancy myself at Rummy" – not be it noted at being a Fellow at twenty-two. The then Provost, Dr. Sheppard (now Sir John Sheppard), had said to Alan, while still an undergraduate, that he hoped he would become a Fellow some day as he was so cheerful, in contrast, the Provost thought, to other mathematicians. We were all curious to see the successful thesis, even though quite above our heads. My husband, reading the opening words of it, exclaimed, "Why! He writes as one having authority." Custom rules that the newly appointed Fellow resumes his former place in Hall among the other B.A.s. Alan rather characteristically overdid this and there were, I gathered, later on, complaints that he preferred the company of the B.A.s to that at the High Table; probably he did not want to assert his position.

Next year, 1936, he submitted for a Smith's Prize his thesis on "The Gaussian Error Function." I found him about to despatch this important dissertation in a most insecure parcel – he never could make up a parcel. He surveyed the package after I had dealt with it, remarking in his comical way, "If I were to receive a parcel like that I should expect it to contain Letters Patent for a Marquisate." Once more this thesis met with success; this time in the award of a Smith's Prize, to the delight of us all and of his old school which celebrated the event with another half-holiday. He had looked forward to the winning of this prize as a certainty and even months before had talked of using it to buy a small sailing boat. He seems to have had the ability to appraise his own work and in later years to assess its worth, yet without any hint of boasting. When I proposed putting together a few notes of his early days for use by a biographer in time to come, he gave an approving grunt. Yet this attitude was quite dissociated from any personal conceit; for instance, he was quite taken aback, when he was about to give a lecture in London, to receive a message of greeting from Bertrand Russell, as he hardly thought he would come within the ken of so eminent a mathematician and philosopher. Again, where he collaborated with others he was most insistent on the credit due to them, while minimising his own share. He always lamented his own slowness, as he deemed it, yet many others were amazed at the rapidity with which he saw the solutions to their problems.

Alan had his own fantastic devices. He used to set his clock by observing when a certain star, viewed from a fixed point, was obscured by a neighbouring building, as this occurred at equal (known) intervals: this method he found quite practical. Indeed he was never at a loss to get round some difficulty in a way entirely his own. Playing tennis on one occasion, in order to remove the puddles on the hard court he borrowed his opponent's hair-grip, with which he made holes by which successfully to drain off the water; and another time, neither of them having a watch, he improvised a sundial.

In 1937 there appeared his paper "On Computable Numbers with an Application to the Entscheidungsproblem,"[1] followed the same year by a short Correction. "Computable Numbers" proved to be his most famous contribution to mathematics. Concerning it *The Times* obituary notice (16th June, 1954) states: "The discovery which will give Turing a permanent place in mathematical logic was made not long after he had graduated. This was his proof that (contrary to the then prevailing view of Hilbert and his school at Göttingen) there are classes of mathematical problems which cannot be solved by any fixed and definite process. The crucial step in his proof was to clarify the notion of a 'definite process,' which he interpreted as 'something that could be done by an automatic machine.' Although other proofs of insolubility were published at about the same time by other authors, the 'Turing machine' has remained the most vivid, and in many ways the most convincing, interpretation of these essentially equivalent theories. The description that he then gave of a 'universal' computing machine was entirely theoretical in purpose, but Turing's strong interest in all kinds of practical experiment made him even then interested in the possibility of actually constructing a machine on these lines."[2] Writing in the Royal Society's Memoir (November, 1955)[3] Professor M.H.A. Newman referred to "Computable Numbers" thus: "It is difficult today to realize how bold an innovation it was to introduce talk about paper tapes and patterns punched in them into discussions of the foundations of mathematics." The following additional comments by Dr. Robin Gandy were published in *Nature:*[4]

During his first years of research he (Turing) worked on a number of subjects, including the theory of numbers and quantum mechanics, and started to build a machine for computing the Riemann Zeta-function,

---

[1] *Proc. London Math. Soc.* **42** (2), 230 (received 28th May, 1936). Correction *Proc. London Math. Soc.* **43** (2), 544.
[2] Quoted by courtesy of the Editor of *The Times.*
[3] M.H.A. Newman (1955). Alan Mathison Turing. 1912–1954. *Biogr. Mems. Fell. R. Soc.* **1**, 246–252.
[4] *Nature* (Messrs. Macmillan & Co., Ltd.), 18th September, 1954.

cutting the gears for it himself. His interest in computing led him to consider just what sort of processes could be carried out by a machine: he described a 'universal' machine which, when supplied with suitable instructions, would imitate the behaviour of any other; and he was thus able to give a precise definition of 'computable,' and to show that there are mathematical problems whose solutions are not computable in this sense. The paper which contains these results is typical of Turing's methods; starting from first principles, and using concrete illustrations, he builds up a general abstract argument.

In the vacations Alan and I had long walks together and whole days out when he would talk about his work and plans for getting down to research into the foundations of mathematics. Certain places round Guildford are associated with "variables" and "constants" and the square root of -1. Though he must have known that I could follow him only in a fog, he liked to share his projects with me even in a very limited measure. Particularly did he endeavour to make me understand at least the drift of his paper on "Computable Numbers," sometimes banging on the ground with his stick to drive home his points. He always presented me with an off-print of his publications, now to prove so useful. My copy of "Computable Numbers" has three kinds of marginal mark – one against what I understood, another to denote what with further explanation I might grasp, while the third mark indicated what no power on earth could make me comprehend. However, despite my defects, we embarked together on a précis in French of "Computable Numbers" for *Comptes Rendues*. After Professor M.H.A. Newman had kindly vetted it and suggested some alterations, a French expert checked it. Unfortunately the man who was to have communicated it vanished to China, while some correspondence relative to the précis was lost in the post; then the war supervened and Alan, as far as I know, heard no more about it. When Alan was assembling his thoughts on "Computable Numbers," Mr. David Champernowne, now Professor Champernowne, spent a few days with us in the summer of 1935 and I recall the amused scepticism with which he discussed the possibility of constructing a machine such as Alan then envisaged to demonstrate his mathematical argument. He mockingly maintained that to house it a building the size of the Albert Hall would be required.

My information on good authority is that "Computable Numbers" contained ideas later incorporated in all general computing machines. It was, I think, this paper which so much impressed Scholz of Münster. It was very well translated into German, with excellent renderings of Alan's original expressions. Scholz sent a photolithoprinted reprint containing the gist of the paper and amused Alan very much with the report that he had *vorgetragt* it among his students. Though so young (twenty-five) Alan was gaining a reputation abroad and before the Second World War much of his work was embodied in the *German Mathematical Encyclopaedia*. Letters from far distant correspondents very often arrived for him, once even from a Japanese mathematician.

In 1937 two other papers – quite short – were published in the *Journal of Symbolic Logic*. These were, "Computability and λ-definability," and, "The p-function in λ-κ-conversion." In answer to a question of mine regarding the application of mathematics to mundane ends Alan referred to something he had been working on, which might be of military value. He gave no details. But as he had some scruples about the application of any such device, he consulted me about its moral aspect.

About 1937 Alan planned to write a paper, "On a theorem of Littlewood": accordingly, with this aim he consulted Dr. J.E. Littlewood, F.R.S., who, though interested, considered that it presented formidable difficulties. Undeterred however, Alan, at some time, carried out his project, but did not publish his paper, lest it should involve encroachment on someone else's special subject. This paper, edited by some other mathematicians, may be included in the volume of Alan's collected works to be published by the North-Holland Publishing Company. It is judged to be a work of singular merit.

At 16 years of age

KING'S COLLEGE,
CAMBRIDGE.

The Rules of GO

i The game is played on a board on which are ruled dly equidistant parallel lines and another nineteen parallel equidistant lines at right angles to them. The intersections of these lines will be called 'places'. Two places are said to be adjacent if the line joining them (not produced in either direction) does not pass through another place.

ii The game is played by two players who are supplied with (an unlimited supply of) counters The players turn by turn put counters in the places of the board. Certain places are inadmissible by rules

Specimen of handwriting

5

At the Graduate College, Princeton

The Farringdon Road market was a favourite haunt of Alan's: it was here that he picked up his violin, already mentioned, and here, I think, found the ancient sextant which was to be part of his equipment for his voyage to New York on 23rd September, 1936. He sailed steerage and I saw him off at Southampton. As we did not realize the extent of the docks we walked from the train to the steamer and the sextant was allotted to me to carry. Of all the ungainly things to hold, commend me to an old-fashioned sextant case. Though some readings were taken, what with the movement of the ship, and a defect in the instrument and Alan's inexperience, he doubted their accuracy. He heads his letter from the Berengaria 41° 20′ N 62° W.

A week after arrival at the Graduate College, Princeton, he writes: "The mathematics department here comes fully up to expectations. There is a great number of the most distinguished mathematicians here – J. v. Neumann, Weyl, Courant, Hardy, Einstein, Lefschetz, as well as lots of small fry. Unfortunately there are not nearly so many logic people as last year." His next letter suggests some disillusionment. "Church had me out to dinner the other night. Considering that the guests were all university people I found the conversation rather disappointing. They seem, from what I can remember of it, to have discussed nothing but the different States they came from. Description of travel and places bores me intensely." On the other hand the graduate students, many of whom were occupied with mathematics, did not mind talking shop – very different from Cambridge in that way. The letters from Princeton are the fullest and best I have from Alan. Fresh impressions, with maybe a touch of nostalgia, encouraged him to write longer and more frequent letters. He comments on tricks of speech and manners. Though prepared to find democracy in full flower, the familiarity of the tradespeople surprised him; he cited as an extreme case the laundry vanman who, while explaining what he would do in response to some request of Alan's, put his arm along Alan's shoulder. "It would be just incredible in England." But that was long before "ducks," etc., were in common use here. At first he liked people saying, "You're welcome," as he thought he was being welcomed. He was evidently unfamiliar with the expression – an Irishism introduced into America, or is it an Americanism brought to Ireland? Anyhow later it used to make him quite apprehensive when it came "back like a ball thrown against a wall."

In his first term at Princeton Alan gave, at Professor Alonzo Church's suggestion, a lecture on Computable Numbers, but was disappointed at the poor attendance. He gathered that unless one had made a name one's lectures were no draw. For a very poor lecture by some celebrity the hall was packed. Here at Princeton he had two things on hand in the Theory of Groups, resulting in one paper "The Extensions of a Group" received by *Compositio Mathematica*, a Dutch periodical, in March 1937 and published in 1938, and the other paper "Finite Approximations to Lie Groups" received by *Annals of Mathematics* in April 1937, which after revision in September 1937, appeared in January 1938.

Apart from work he enjoyed play-reading and was a member of a college play-reading society which met most Sunday evenings. He must have been fairly good as he was also at one time a member of a rather exclusive play-reading society or club at King's. Almost at once he played hockey for the Graduate College and found it "great fun," though later he was involved in making arrangements for matches, etc., to a greater extent than he quite wanted. They played various other teams at Vassar, New York and doubtless in other places. "Thanksgiving" he spent with someone in New

York who had been asked to befriend him. He amused himself in exploration of the less fashionable parts of the city. At the New Year (1937) he went ski-ing in New Hampshire with Mr. Maurice Pryce (now Professor Pryce) and Mr. Wannier, visiting Boston on the way back. He wrote most appreciatively of Mr. Pryce's kindness to him, and of how he had welcomed him on his arrival at the Graduate College and been such good company. In June he went with Mr. Pryce to stay with a cousin of mine in Rhode Island. In this cousin he found a kindred soul, for he was an amateur astronomer and had a little observatory with a telescope made by himself, for which he had ground the mirrors. In fact Alan described him as a competitor for the Relations Merit Diploma, a distinction of Alan's invention. A letter from my cousin testified to reciprocated admiration.

Less than six months after Alan's arrival at Princeton, Dean Eisenhart, Dean of the Graduate College, and Mrs. Eisenhart, began to press him to pass a second year at Princeton. To quote Alan: "Mrs. Eisenhart put forward social or semi-moral, semi-sociological reasons why it would be a good thing to have a second year. The Dean weighed in with hints that the Procter Fellowship was mine for the asking (this is worth $2,000 p.a.)." He was rather hesitant about assent to the suggestion, partly because most of his friends were leaving Princeton, and partly because he was uncertain how two years' absence from King's would be regarded. Though Alan himself considered that a poor reception had been given at Princeton to "Computable Numbers," Dean Eisenhart does not seem to have shared this view for he wrote to me towards the end of Alan's first year at the Graduate College: "We have enjoyed having him here very much, not only for his personal qualifications but for the impression he has made on our men in mathematics."

Second Year at Princeton

After the summer vacation of 1937 spent in England Alan returned to Princeton as a Jane Eliza Procter Visiting Fellow. The time on the voyage was whiled away by philosophical discussions with Mr. Will Jones and by position-finding with the aid of his famous sextant. Back at the Graduate College he started work on his dissertation for his Ph.D., the subject being "Systems of Logic based on Ordinals." He received the degree in May 1938. The thesis had been delayed, partly because Professor Alonzo Church made suggestions which resulted in its expansion to greater length than at first envisaged, and partly because Alan was unfortunate in his typist. He longed for the co-operation of Miss Pate in Cambridge. In a footnote to "Systems of Logic based on Ordinals" he acknowledges his debt to Professor Alonzo Church for most valuable advice and assistance. This paper was published in 1939 by the London Mathematical Society in Ser. 2, Vol. 45. According to the Memoir in the Annual Report of the Council of King's College, Cambridge, November 1954, this "work on ordinal logics which he wrote for his Doctor's Degree at Princeton and a notable paper on the word-problem, written towards the end of his life, have been considered by some good judges at least equal, in power and originality of thought, to the more famous one on computable numbers."

In the Christmas vacation he went with Mr. Martin to stay with him in South Carolina and on the way back visited a Mrs. Welbourne and her family who proved very agreeable. Mrs. Welbourne used to get from the Dean of the Graduate College the names of visiting English students whom she enjoyed entertaining in this way. In the Easter vacation he and Mr. Will Jones went to Washington; they visited the Senate and were struck by its informality. At St. John's College, Annapolis, where they lunched, they saw something of the methods employed there.

My husband advised Alan to find a job in America. Alan himself, though not very keen, made tentative enquiries, but eventually refused the offer of a post as von Neumann's assistant. Accordingly he was back in England in the summer of 1938 and settled down to research and the academic routine at King's, where his Fellowship had been renewed. This was the kind of life he then most enjoyed, interspersed with vacations in Guildford. Vociferous singing of hymns proceeded from his study at home and appeared to indicate that research was going well. I used to complain of his unseasonal choice of hymns – Easter hymns at Christmas and *vice versa*, Lenten ones at festivals. Once more in Cambridge he was probably engaged on his paper, "A method for the calculation of the zeta-function" which was received by the London Mathematical Society on 7th March, 1939, but did not appear in their *Proceedings* until four years later.

Two other subjects occupied his attention about this time: with M.H.A. Newman he produced a paper entitled "A Formal Theorem in Church's Theory of Types" published by the *Journal of Symbolic Logic* in 1942. The same journal published his paper on "The use of dots as brackets in Church's system," also in 1942. In the Lent term 1939 he gave a set of lectures and thought they were going off rather well, though he quite expected that attendance would drop off as the term advanced. Otherwise this period at King's seems to have been uneventful.

6

Some Characteristics

Alan was broad, strongly built and tall, with a square, determined jaw and unruly brown hair. His deep-set, clear blue eyes were his most remarkable feature. The short, slightly retroussé nose and humorous lines of his mouth gave him a youthful – sometimes a childlike – appearance. So much so that in his late thirties he was still at times mistaken for an undergraduate; hence occasional attempts were made to "prog"[1] him.

In dress and habits he tended to be slovenly. His hair was usually too long, with an overhanging lock which he would toss back with a jerk of his head. The first thing to be done when he came home was to send him to have his hair cut. When he did take the trouble to comb it, five minutes later he would run his fingers through it so that once more it would be standing on end. At King's, for a time he took more trouble with his clothes and even subdued his hair. But when the war came he was seldom at home, clothing and sartorial advice were both rationed, and he relapsed into the old ways. In his last years there was again some slight improvement and he used to object to complaints about his clothes. The real trouble, perhaps, was not so much the clothes that he wore as the way in which he wore them. Considering how little his appearance indicated his academic stature and how little he did to advance his own interests, it was a matter of wonder to me that wherever he went he was quickly recognized as a remarkable person: he

---

[1] "Prog" (undergraduate slang) refers to disciplinary action taken by Proctors who found undergraduates out after dark not wearing their gowns.

admitted that he was negligent in applying for posts or putting himself forward.

He could be abstracted and dreamy, absorbed in his own thoughts which on occasion made him seem unsociable; this was partly because he had no gift for "small talk" unless spiced with interest or humour. The ability of anyone else to persevere in it amazed him. Once some guests proved heavy in hand and failed to develop topics of conversation. After they had left, Alan exclaimed: "Mother! I don't know how you do it. For the last two hours you've been working with rope and pickaxe."

Despite his abstraction he gave many people the impression of a vivid, lovable personality – indeed "vivid, lovable, modest," were the epithets constantly employed by his friends in writing of him. He showed a modesty regarding his personal achievements which did not preclude him from being emphatic and assured as to facts and how work should be done. No doubt he was a strange mixture of realist and visionary. Indeed he surmised that the seclusion of a mediaeval monastery would have suited him very well.

But Alan's head was not always in the clouds. He took a real interest in his fellow creatures. Though he had no more than the salary of a Fellow, just before the Second World War he made himself responsible for all the expenses, except school fees, of an Austrian refugee of fifteen. As self-constituted guardian of the boy, Alan paid for all his clothing, holidays and extras, and later paid all his expenses at Manchester University. The boy became a naturalized British subject and took the name of Robert Arnfield. He very gratefully acknowledged his debt to Alan for his advantages in life and made the most of them.

Many friends were impressed by Alan's courage, and several of them stressed quite independently his almost fanatical integrity. In general his disposition was a cheerful one and he was blessed with the ability to laugh at himself. At times however, he would get annoyed by crass stupidity. Any loose use by me of a scientific term, such as my calling water-vapour from a boiling kettle "steam" always met with a disapproving correction. For unaccountable reasons he would be put out and then, without explanation, would depart and walk off his mood. What made him very angry indeed was to be contradicted on scientific points. He was, moreover, intolerant of authority not justified by ability. This may have rendered him a difficult subordinate. One curious thing was that, though he was particularly candid with me, sometimes an apparently innocuous comment or question made him retreat suddenly into his shell. He admitted that he was liable to quarrel when he went on holiday with only one companion: in retrospect he could see how absurd it was, and once, with rather shamefaced amusement, narrated how he and a friend, after some trifling dispute, had pursued their way a quarter of a mile apart along a deserted straight French road.

From Sir Geoffrey Jefferson, F.R.S. came this vivid impression: "He was so unversed in worldly ways, so childlike it seems to me, so unconventional, so non-conformist to the general pattern . . . so very absentminded. His genius flared because he had never quite grown up. He was, I suppose, a sort of scientific Shelley." Though to a large extent Alan ignored accepted conventions, paradoxically he adhered rigidly to some. Thus he strongly objected to being addressed as "Doctor" in this country, since his Doctor's degree was an American one and therefore, in his opinion, incorrectly used outside the United States of America. Some of his young friends, aware of these scruples, enjoyed teasing him by using the title.

He was very shy and though some may not have liked him the less for that, there were times when his shyness led him into extreme gaucherie. To some his unconventionality, coupled as it was with complete absence of "stuffiness," was refreshing, while to others it proved perplexing, if not even objectionable. He, himself, was concerned and sensitive about being eccentric, yet found no remedy, for he was unable to escape from his innate originality. Indeed he may have been dimly aware that, as the foreword hints, he was not a man of his own time: a fellow mathematician summed him up as "the rough prototype of the coming age of Science and Machines."

Shyness made him a slow starter in making friends, but once the ice was broken he had the power to win affection and could claim a wide circle of friends, as the King's College Annual Report, November, 1954, testifies: "Although in later years his work lay away from Cambridge, he continued to visit the College from time to time, where his wit and resource in argument, his gift of humour and his modesty had endeared him to very many friends." He was always ready to help others less gifted, as when, though busy with his own work, he read some psychology in order to assist a friend in difficulties with this subject. His ability, zest, and appreciation of the difficulties of others are thus summarized by Flight-Lieutenant Ivor Jones:

> Alan Turing had the ability to absorb a page of mathematical argument with the quick glance that lesser mortals reserve for newspaper paragraphs, and he would plunge ahead developing the argument mentally, pronouncing theorems and consequences that lay many pages of calculation away. I have never met a mathematician who was so completely the master of all branches of his subject. Above all he was *enthusiastic* – he communicated his enjoyment to his students and colleagues and made their work seem an exciting challenge rather than the solemn ritual so many instructors made of it. When one confessed to difficulties, he was a patient and sympathetic guide and I remember with gratitude his help and encouragement.

Yet despite his profound mathematical insight, eagerness to press on often resulted in elementary mistakes in the simplest arithmetical calculations. Indeed he illustrated in his own person the point he made in his paper "Intelligent Machinery. A Heretical Theory", where he stated: "I believe that this danger of the mathematician making mistakes is an unavoidable corollary of his power of sometimes hitting upon an entirely new method. This seems to be confirmed by the well-known fact that the most reliable people will not usually hit upon new methods."

He was very happy-go-lucky in various ways: once when dining with friends in London he left his bicycle out in the street, heedless of the warnings of his fellow diners. Needless to say, by the time they came out the bicycle had disappeared. No count remains of the number of bicycles, waterproofs and other possessions that he lost. He was liable to be "tied up with string," and put great faith in it. One of his friends once found Alan's bicycle, with an accessory motor, all tied up with string – and this for use in the rapid and heavy traffic of Manchester. In protest the friend disabled the bicycle by cutting these makeshift repairs.

Absent-minded people tend to be unpunctual but Alan had a marked sense of time and was very punctual; even on long cross-country runs he almost always gauged the time so as to be home punctually. Any waste of time caused by unpunctuality in others irritated him. However, this evaluation of time did not mean that he was constantly at work, for he could relax completely, and as relaxation he systematically took outdoor exercise of some sort. From childhood he had been given to taking up crazes of one kind and another, on which in turn he would concentrate intensively, e.g., making paper boats and kettles, playing chess, gardening or doing scientific experiments.

It is hard to think of Alan without remembrance of his boisterous resonant laugh that came with such sudden outburst from one normally so quiet. When Dr. Robin Gandy and I were his fellow guests, great peals of laughter proceeded from the kitchen as he and Robin combined cooking with investigations into *hydrae* just collected from a murky pond. One friend has described Alan as a "good enjoyer of jokes, rather than an originator of jokes," but I would say that he did not so much crack jokes as describe incidents in a strikingly humorous and original way. In serious discussion his gift for comic analogy delighted many people. It amused him to picture himself in some fanciful situation. Thus, as he was setting out one day armed with flowers to visit an elderly lady, he remarked that he thought of visiting a dancing-master *en route* in order to learn the appropriate bow with which to present his bouquet. Professor Fred Clayton recalls the following typical incident:

Shortly before he was due to talk on the Third Programme, Alan and David Champernowne concocted with a recording apparatus a bogus discussion which they tried to persuade me was the real thing on the wireless. As I knew about the broadcast, without knowing its date, I was the more easily deceived. Alan played up very well, and for the first few minutes, though somewhat perplexed, I did swallow it, until some of the other speakers' remarks – they were all played by David, I think – strained my gullibility to breaking point.

The following reminiscence, also coming from Professor Clayton, shows other aspects:

I hadn't much to do with Alan in our first two years at Cambridge, though I coxed a boat in which he rowed. I think I classed him as the more taciturn type of mathematician . . . It was comparatively late in the day that I realized that he was not only brilliant in his own line but had much wider interests than I imagined. He always took me by surprise whether with Fellowships or cross-country running. I owe him two very happy holidays and some unselfish and brilliant assistance with a piece of research that can have had little intrinsic interest for him. The first of these holidays was at an unpromising enough time, being just before the outbreak of war in 1939. We had all been together in Cambridge – Alan, myself and the two refugee boys, Karl and Robert . . . Alan was taking the two boys on a sailing holiday anyway . . . I felt I might just – as well go and join them . . . Alan had to do all the sailing, and the three of us were all more or less idle and in the way. He was very patient and initiated us all into the mysteries of sailing. I enjoyed being crew, and was very content to leave the complications and responsibility to him. Perhaps it was partly the feeling that war was imminent that made this holiday so memorable and enjoyable.

We lost contact almost completely during the war, but after it, in 1947, I suddenly felt I would like to repeat this very happy experience . . . He fixed everything up. We hadn't the boys with us, but the tail-end of a glorious summer and no threat of an imminent war. It was different, but certainly not disappointing, as repetitions of this kind so often are. I particularly remember a very satisfactory sail in perfect weather right over to the Isle of Wight.

Later, when I was already in Exeter, I unashamedly picked his brains on certain problems of ancient astronomy that I had become involved in. Here, as once or twice before, I found the contrast between our methods of approach amusing and revealing. I fear he found both me and the ancients equally obtuse at times from the point of view of a modern mathematician. I remember trying to persuade him that, on a particular matter, the ancient way of looking at things, however inexact and erroneous, was the most natural approach at a more primitive stage. It was, indeed, the stage I was still at, but he seemed sure that no one, not even as a child, could have been tempted to such a curious view. And I was suddenly reminded of an incident on our second sailing holiday when there was some confusion about ration books. He saw the whole thing so quickly and clearly that he could not see why the hotel manager, or whoever it was, could not do likewise. On this occasion also I found myself in sympathy with the more primitive mentality, and trying to put the brake on and interpret.

He was never as much interested in either literature or politics as I was. But he could make remarks on such topics from his own point of view that were interesting partly because of this detachment. Once he thought I sounded unduly shocked or surprised at German scientists working happily for Hitler and argued that it was natural to all of us academic researchers to be absorbed in the job and to think little of its implications. But he did concede in the end that such detachment might have been more difficult in the Reich. On another occasion he suggested that our electoral system might soon become quite absurd and unworkable because, with shifts of population and fewer safe seats for either party, it would become more and more possible for a government to get elected with a large majority in the House and no majority at all in the country. I felt that proportional representation naturally appealed to him as more mathematically logical. But he had some ingenious way out of the dilemma by casting lots or something of the kind under certain circumstances. I was interested at the idea of this old Greco–Roman device returning into politics. His system was based on the laws of probability and, he argued, perfectly fair, adding 'but the politicians would take a lot of convincing, I suppose.'

Once when I was at the helm and commented on the delayed reaction of the rudder that tempted one to over-do every change of course, he said he was convinced there was a sort of law of over-compensation which had a very wide application in various fields, and he proceeded to give illustrations. It was one of many ideas that he felt he might go further into some day. One never knew what he might take up next.

It is difficult to account for Alan's stammer. In broadcasts there was not a trace of stammering when he had the script before him and knew what he was going to say. My own theory is that thoughts and modes of expressing them tumbled over each other in his mind too quickly for him to give voice to them in dearly articulated phrases. But he seemed quite unembarrassed by this occasional failing. One of his friends remarked, "Alan has the most uninhibited stammer I have ever heard." He turned a deaf ear to the suggestion that to improve his delivery when lecturing he should take lessons in elocution. Strangely incompatible with his sturdy physique – and he could run twenty-six miles at a stretch and end up smiling – was his tendency ever since childhood to faint, especially in church and school chapel. Once having got over-tired on a walking tour he had a narrow escape when he fainted in his bath. Medical examination could find no physical cause for this idiosyncrasy, which he never outgrew and about which he tended to be apprehensive. For this reason he could tolerate accounts of neither surgical operations nor accidents. When guests embarked on graphic descriptions of them, and Alan began to turn green, I had hastily to despatch him on some trumped-up errand, to get him out of the room. He understood the tactics, and did not have to search for the non-existent letter he was asked to post.

One thing that Alan specialized in was his choice of presents. Not only were they on a very generous scale but great thought and care went into finding the right gift to suit the age and taste of the recipient. Whereas most of us, I imagine, after consideration of the recipient, rack our brains to think of the right present and then shop, his system worked in reverse order. Some time before Christmas he made a reconnaissance of the shops to see what was available and then decided what was appropriate for each person. He admitted that he used to get quite worn out with searching and deciding. Some of the things in most constant use in my house were his gifts. His present to one small niece called forth the spontaneous exclamation, "What a Christmas!"

Other examples of his generosity come to mind. He was quick to respond to the "Week's Good Cause Appeal" and would often get out his cheque-book before the speaker had finished. The cheque would be posted with no covering letter. On visiting a favourite aunt he found she was losing her sight, so he at once invested in three sets of Braille apparatus – one for her, and the other two for the writing of letters to her by Alan and myself. He started to master writing Braille and instructed me. Unfortunately his aunt's powers of concentration were too much impaired to profit by his kindness. Again there was the case of the friend who had given up her much loved mission-work in Central Africa in order to look after her mother. On the death of her mother, of whom Alan was very fond, he immediately proposed that his friend should revisit Africa and he offered (and eventually gave) financial help towards her return there. These instances of thoughtfulness reveal the sensitivity hidden beneath his somewhat rough-hewn exterior. Only those who knew him best discerned that underneath he was, in some respects, very sensitive.

Taken all round he presents a strange study in light and shade. To some people he appeared baffling and complex, while to others, including myself, he displayed, in marked contrast to his great intellectual gifts and achievements, an unaffected simplicity. It was surprising that, no matter how much absorbed he was in his work he never seemed to mind being interrupted. This may have been because his whole train of thought was clear in his brain. When he gave a lecture at Sherborne School, a former master expressed surprise that he did not wear spectacles considering all the calculations he had to do, to which Alan replied: "Oh! I do them all in my head." Similarly he memorized dates and times of all his appointments, for as far as I know he never kept an engagement book.

It must have been the childish streak in Alan which made him so much liked by and at home with children. It was always with enthusiasm that he helped me to organize games and competitions at my Christmas parties for children and their parents. He loved to "baby-sit" for the little boy next door; he would set his minute-timer for this agreeable task and go upstairs at half-hourly intervals to listen and assure himself that all was well. A friend who spent a cycling holiday with Alan in France told me how in the French shops children paid no attention to him, but gathered round Alan, whose attitude towards children was that of "man to man" and was founded on a sympathetic understanding of them. He would take great pains for them. Thus for one of his very young friends going abroad he wrote out a method of playing Solitaire to amuse her on the journey. His instructions include three diagrams; writing to this little girl of seven, as usual he raises her to his level and says: "I find it helps, if I am trying to do the puzzle to use four kinds of pieces like this (see diagram) or better still to use a board with the squares in four colours. Each piece always stays on the same colour until it is taken." He then continues with further warnings and advice. The letter was carefully posted "Express" – this from one who normally shirked letter-writing and caused exasperation by his dilatoriness over correspondence. Professor Newman's sons were among Alan's young friends. One of them, as a small boy, when asked who were coming to his birthday party, said with pride, "Six boys from the school and a bachelor" – the bachelor was Alan.

Later, at Wilmslow, Cheshire, Alan was in frequent request by the little boy of four who lived next door. At a call from this child he would at once set aside his work and let himself be pummelled, or join in a game or discussion. The roof of the garage was the spot which they found favourable for conversation. When the boy's mother broke in on their last discussion she learnt that they were debating whether God would catch cold if He sat on the damp grass. How valuable a tape-recorder would have been for preserving their views on this subject – one worthy of the schoolmen.

7

War Work in the Foreign Office

We had given up our home in Guildford in March 1939, and being mostly on the move until war broke out we saw little of Alan: during the next six years he could spend only snatches of leave with us, for immediately on the declaration of war he was taken on as a temporary Civil Servant in the Foreign Office, in the Department of Communications.

As a "back-room boy" he was not allowed to enlist, though he served for a time in the Home Guard. At first even his whereabouts were kept secret, but later it was divulged that he was working at Bletchley Park, Bletchley. No hint was ever given of the nature of his secret work, nor has it ever been revealed. The enforced silence concerning his work quite ruined him as a correspondent: his letters from then on became infrequent and scrappy. However on his occasional visits to us and on my visits to him I heard of some of the truly Alanesque things that happened during the war years.

His ability was early recognized in his department where he became known as "the Prof.," or simply "Prof." He lived at the Crown Inn, Shenley-Brook-End, some three miles out of Bletchley. Here his kind landlady, Mrs. Ramshaw, took great care of him and generally mothered him and admonished him about his clothes. Someone on the staff at Bletchley Park reported to a relative of ours that Alan was "wrapped up in his theories and wild as to hair and clothes and conventions, but a dear fellow." Alan himself deplored the shabby clothes of some other people at Bletchley Park and complained that they wore them "not even patched." It was, of course, the time when clothing coupons restricted outlay on clothes.

Alan used to wander round the fields in the neighbourhood and evidently his peering into hedges and ditches in his inquisitive way made him appear a suspicious character to a local spy-hunter. Once when he happened to take the same walk two days in succession he found a couple of policemen in readiness for him on the second day. He was able to produce an identity card, unsigned however, because as he maintained, we had been told at some stage to write nothing on our identity cards. Some questioning elicited the fact that he was employed at the Foreign Office, whereupon the investigation broke down. The idea of "Prof." being nearly arrested caused much amusement in his department. The extent to which "Prof." took precedence over his own name was borne in on me once when I was with him and had to telephone to his department explaining his absence, due to influenza. Though I got through to the right number and extension no one seemed to have heard of "Mr. Turing." It was pointed out later by one of his assistants that to have referred to him as "Prof." would have avoided all uncertainty.

In the shelter during air-raids he knitted himself a pair of gloves, with no pattern to guide him, just out of his head; he was, however, defeated when it came to completion of the fingers, so he used to bicycle in from Shenley with little tails of wool dangling from his finger tips until one of the girls in his office took pity on him and closed up the ends. He found bicycling in the neighbourhood of Bletchley the worst possible thing for his acute hay fever, but he discovered that his gas mask gave him some protection from the pollen, and used this discovery without hesitation. As more than one friend has said, he was completely indifferent to the opinion of those who might deem his behaviour silly, "judging, as he did, everything on its objective merits, as he saw them." For him the logical course was of more account than the conventional.

Professor Newman has already mentioned in the Royal Society's Biographical Memoir the famous Bletchley bicycle with its insecure

chain. Many colleagues have commented on his original method
of dealing with its defect. He worked out that the chain came off
after x revolutions of the pedals and at first counted these revolutions so as to be ready to execute the necessary manoeuvre in
order to keep the chain on. This was tedious, so he fixed a little
counter, on the bicycle to warn him of approaching trouble. When
investigations were carried further, he discovered the mathematical relation between the number of pedal revolutions, links in
the chain, and spokes in the wheel. This showed that the chain
came off when a slightly damaged link came in contact with a
certain bent spoke. The spoke was straightened. This done, there
was no more need to bring a bottle of turpentine and piece of rag
to the office to clean his hands after replacing the chain whenever it did come off. A bicycle mechanic would have fixed it in
five minutes. The incident is trivial but as Mr. A.C. Chamberlain
writes:

[I]t illustrates three of Alan's characteristics as I saw them:

(1) His indifference to other people's scorn or jokes. Most people would
    have kept the turpentine and rag in the saddle bag so that the whole
    office need not know about it.
(2) The determination to find out and rectify the fault himself.
(3) The method of going about it by observation and theory.

Early in the war he realized that should Britain come under
German occupation bank accounts would be useless. He therefore
bought two ingots of silver bullion which he buried in two different
spots, counting on recovery of the silver when the Germans should
be driven out. He used to tell this story rather shamefacedly, though
it was quite a sensible proceeding, had it been efficiently carried
out. He conveyed the ingots in an ancient perambulator and in
lifting the heavy bars slipped a disc. Though he retained a cryptic
plan of the hiding places he failed to discover the silver later, when
he used his home-made mine-detector. Mr. Donald Michie helped
him, and in return for his assistance, was offered a percentage

of the value of the hidden treasure, or his travelling and hotel expenses plus £5 for each expedition. He wisely chose the latter alternative. They set out armed with brown paper, spades and the perambulator to one site, but because the mine-detector did not prove strong enough and Alan was unsure of his landmarks, the ingot buried at that site was not found nor anything further heard of that buried at another spot.

Mr. Michie classed Alan with five other men whom he considered as being at the "genius" level. He maintains,

all these men have another thing in common which perhaps goes hand in hand with great intellect – namely, an unflagging boyish gusto over any project or topic which is raised. This was particularly so of Alan, and his enthusiasm over the treasure-hunting and chess-machine provides a good example. When I first met Alan his eccentric manner deceived me into thinking he was all head and no heart. When I knew him better I realized that his emotions were so child-like and fundamentally good as to make him a very vulnerable person in a world so largely populated by self-seekers.

The most that Alan told me about his war work was that he had about a hundred girls under him. We knew one of these "slaves" as he called them. From her came the information that they marvelled at her temerity in greeting him on Christmas morning with "A Happy Christmas, Alan," for they held him in great awe, largely because when he rushed into their part of the building on business, he never gave the least indication that he even noticed them. The truth probably is that he was equally alarmed by them. When he was on the night shift for a fortnight at a time, he would take his weekly day off in the middle of the period and very sensibly without upsetting the routine of daytime sleep he would appear at the office about midnight to spend, with access to meals, his free night on his private mathematical research. He would sometimes work at home out of hours on some abstract theoretical office problem that was particularly absorbing, but not of a secret nature. He was not much of a newspaper reader: so engrossed was he in his duties that he did not know of the invasion of Norway until he had to do something about it in his official capacity. Fairly early in the war, about 1942 maybe, he and two colleagues were summoned to Whitehall, thanked for their work, awarded £200 each free of tax and given the use of a car for the day. In the summer of 1941 he snatched a few days' holiday at Portmadoc in Wales, and with a friend did some of the climbs we had done before, but from other points.

One of Alan's colleagues who had worked with him at Bletchley wrote to me years later:

Alan was of course in a class of his own in the sphere of brain power but what impressed us all was his championing of the underdog and his willingness to help others who had problems which to him were just simplicity itself. I still have in my possession some corrections he made to a problem I was engaged on, and the occasions when he threw out an immediate solution are too numerous to mention.

In September 1942 he was told to keep himself in readiness to go to the United States, though actually he did not sail until November. He had on arrival some difficulty over admission as he had been told on no account to take any papers other than those in the Diplomatic Bag which he carried. The triumvirate who confronted him on landing talked of despatching him to Ellis Island. Alan's laconic comment was, "That will teach my employers to furnish me with better credentials." After further deliberation and passing of slips of paper, two of the triumvirate outvoted the third member and he was admitted. Even on Ellis Island he would have found something of interest, perhaps more than he found at Washington. It was not the only occasion on which he was sent on a secret mission without proper arrangements being made. For instance, once when ordered abroad he asked about money and was assured he would be met on the other side and all necessary expenses paid there. Through some failure of administrative machinery this proved not to be the case, but fortunately a friend had given him a few francs, enough barely to pay his taxi on arrival, but not enough to pay for a meal too. Still hauling the diplomatic bag and very hungry, he presented himself at his destination and was questioned by officials who professed to know nothing at all about him. It may have been a routine method of checking his reaction, but he found it a most disconcerting and unpleasant experience which he never forgot.

The voyage to America of course had considerable hazards and for me, some anxiety, but he circumvented the censor by sending me from New York a cable of good wishes for my birthday, which assured me of his safe arrival. The journey, except for "splendid food," was rather uncomfortable owing to overcrowding. He was the only civilian on board apart from a couple of children, and there were nine men to a cabin and at times about six hundred men in the Officers' Lounge, which almost brought him to fainting point. He was able to take advantage of the visit to make sundry purchases, including a "Go" board, and was proposing to attend the "Go" club meetings in his neighbourhood to discover the American standard of play. Washington proved outrageously expensive, almost beyond the limit of his allowance. There was some hold-up about his job, which involved a useless period of idling in New York. Though he did some desultory mathematics of his own, the atmosphere was not conducive to study.

He remained in America until March 1943. He seems to have taken the opportunity to visit Princeton and probably saw something of the progress of computing machinery in the States. He returned in a destroyer or similar naval vessel and experienced a good tossing on the Atlantic. He was, however, fully occupied, for a colleague tells me that on the voyage home he worked on his plan for a speech secrecy device. He adds:

His only book was a twenty-five cent publication on valve characteristics. With his superb brain he worked out a device which worked and was, I think, years ahead of its time.[1] I always remember his definition of a device; it was something like this. 'Give the enemy the circuit, all the components, let them make it and then it must be impossible for them to break down the message.'

A colleague at Bletchley, who proved a most staunch friend, sums up Alan's work there thus:

In our work as everywhere else his profound originality was his most striking characteristic; he was quite bad at routine work (whenever he had to do it). He was always impatient of pompousness or officialdom of any kind – indeed it was incomprehensible to him; authority to him was based solely on reason and the only grounds for being in charge was that you had a better grasp of the subject involved than anyone else. He found unreasonableness in others very hard to cope with because he found it very hard to believe that other people weren't all prepared to listen to reason; thus a practical weakness in him in the office was that he wouldn't suffer fools or humbugs as gladly as one sometimes has to.

According to Dr. I.J. Good, he sometimes offended the administrators.

'Prof.,' had an impish sense of humour, some of which was directed at authority. On one occasion he ordered a barrel of beer to be sent to the office, but it was disallowed. I think he ordered the beer partly because he knew it was unconventional to have beer in the office, and not because he was the sort of man that one envisaged with a tankard of ale in his hand.

From the official point of view he may not have been amenable – officialdom and genius make reciprocally uncongenial company.

When Alan's task at Bletchley was satisfactorily completed, he was transferred to a different establishment for other duties calling for the exercise of his special talent for original work. Mr. A.C. Chamberlain writes:

---

[1] Some of my father's inventions were likewise said to be far ahead of his time—in fact scoffed at at first and long afterwards adopted.

I used to visit him there occasionally, and I do not think anything showed his character better than the way he got to work there. He was quite on his own and had to do the wiring-up of the apparatus he designed himself. His accommodation was an old cottage . . . He nevertheless set to work quite happily to do the job . . . Alan was easily the most brilliant man I have met.

A friend also employed at this place writes: "I well remember Alan's class for the 'lower orders' when in the evenings he gave a series of lectures on valve theory . . . His great gift of simplification of difficult problems was very popular." There seems to be some difference of opinion about his ability to put things clearly. Professor Newman describes him as a difficult author to read; while Dr. Gandy and others emphasise his gift for simplification. Possibly in treating a matter orally he could sense how far he was making himself clear and adapt his methods accordingly: this however would not apply to broadcasts where he succeeded in making his points intelligible to the average hearer. All I know of this time is the story he told me of how some men in a non-operational unit with which he was working invited him to a secret party at which none of the guests was to be of higher rank than lance-corporal. That the King's College don, in their eyes, ranked no higher than a lance-corporal is an indication of the simplicity of Alan's behaviour and lack of "side." Actually there was some hitch and the secret party never came off. But the invitation gratified him, for he liked these men and respected their intelligence. A friend of mine, who had herself been headmistress of a large High School, has never forgotten how he came, once as a schoolboy and once as a Fellow of King's College, to give two scientific talks to a group of women at the Adult School in Guildford and how he did not think it beneath him to share his knowledge with people of a very different mental calibre. It was all very simply done with no suggestion of condescension, and his talks provoked many questions.

Though he had done a certain amount of running for exercise, towards the end of the war he took it up more seriously: incidentally he fell and broke his ankle with apparently no lasting ill effects. Emerging from a "back room" he caused a memorable sensation by winning the mile race in the Regimental sports. The officer who commanded the unit thus enlarges on the incident:

> I was informed that a Professor Turing was to be attached to us for a while for special duties and I arranged for his accommodation both in the mess and also in a hut. In spite of having to live in a mess and with soldiers, Turing soon settled down and became 'one of us' in every sense; always rather quiet but ever ready to discuss his work even with an ignoramus like myself. I was entirely on the Administration side and in no way technical. I well remember when we were arranging some sports in connection with a fête we were running, officers were asked to put up their names for any of the races. Imagine our surprise when Turing put his name up for the mile. We rather thought it might be a leg-pull, but on the Day 'The Prof.,' by which name Turing was also known, came in a very easy first. He left us to work on a new Electronic Brain, and we saw him no more, but all were very sorry to part with such a pleasant messmate who was liked and respected by all ranks.

This ability to win respect in utterly diverse environments is often stressed – at his public school, in academic circles, at his athletic club – and here, (though there was nothing of the soldier in his make-up), among military men. For his part, he always accepted people on their personal merits, irrespective of their background.

Another colleague at the Foreign Office with whom Alan had frequent lunch-time strolls "just marvelled at the depths of Alan's knowledge in so many fields. It did not really matter which subject was under review, for he was able to analyse the problem and give a solution of so many aspects of our job."

Occasionally during the war Alan came to stay with me for, a week or so in Guildford, but owing to the secrecy of his work there is little to report. His chessmen had been stolen during his absence in America and, as it was very difficult to get chessmen, he turned as usual to material at hand and made some very rough chessmen in clay, and, I gather, managed in some improvised fashion to fire them in a tin over his open fire.

For his war service he was made an Officer of the British Empire, as he was one of a team whose joint work was an important factor in our winning the war. Some friends dining with him some years later wanted a nail or screw for something. Alan produced a box in which were odds and ends, nails, etc., and his O.B.E. medal all mixed up together, much to the amusement of his guests. Because of the King's ill-health he did not receive the award personally, but by post, accompanied by a letter. Notwithstanding his unconventionality he had a certain sense of occasion and would have received the decoration with due decorum. I remember well the unexpectedly solemn dignity with which he received his degree, unaware of an errant wisp of hair projecting at the back of his head.

After the conclusion of the war in Europe he was sent abroad on duty for the Foreign Office. One of the other members of the mission recounts: "The last meal of the day was at 5 p.m. and the first at 7 a.m. After this evening meal Turing was observed to take a thick slice of bread, to coat it liberally with butter and peanut butter, and solemnly and slowly to chew it. On seeing our somewhat enquiring glances he explained that he was not hungry then, but thought he would be by 7 a.m. the next morning." As usual he adopted the practical in preference to the conventional course. Behaviour of this kind was probably responsible for the circulation of sundry tales dismissed by Alan and others as purely legendary.

8

At the National Physical Laboratory, Teddington

On release from his duties at the Foreign Office Alan was offered a Cambridge University lectureship, which he declined since his attention was becoming focussed on computing machinery. Before the war he had already begun to build a computer of his own with wider scope, he hoped, than those then in operation. It was natural, therefore, that his aim should be to see his logical theory of a universal machine, previously set out in his paper "Computable Numbers" (published in 1937), take concrete form in an actual machine. On submission to the Government of the outline of his design for such a machine he was taken on to the staff of the National Physical Laboratory at Teddington, and became a permanent member of the Scientific Civil Service in October, 1945. When this service was reconstructed at the end of the war, senior posts were made available for people of exceptional merit so that "they would be able to advance fairly high in the scientific hierarchy without the usual requirement of undertaking considerable organizational responsibility." Under this scheme Alan was one of the first to become a Senior Principal Scientific Officer. In the Mathematics Division of the National Physical Laboratory his task was to plan more fully the logical design of an automatic computing engine. According to *The Times* of 16th June, 1954, "he threw himself into the work with enthusiasm, thoroughly enjoying the alternation of abstract questions of design with practical engineering."

Here he worked in collaboration with engineers and experts in the field of electronics, while the whole team made certain modifications in his design as plans for the construction of the

Automatic Calculating Engine progressed. By November 1946, sufficient advance had been made for news Of this venture to be broadcast by Sir Charles Darwin, F.R.S., then the Director of the National Physical Laboratory.[1]

The project has been picturesquely called the Electronic Brain. For a long time mathematicians have been occupied in getting better logical foundations for their subject, and in this field, about twelve years ago, a young Cambridge mathematician, by name Turing, wrote a paper which appeared in one of the mathematical journals, in which he worked out by strict logical principles how far a machine could be imagined which would imitate processes of thought. For example, the cash register in a shop, whenever you feed into it two and then two, will hand out the answer four, and arithmetic machines can automatically do much more complicated things than that. That part is easy, but Turing set himself to find out the ultimate limitations that such a machine must have. The answer cannot be given simply, of course, but it is roughly that you could make a machine do anything which can be called rule-of-thumb.

It was an idealized machine he was considering, and at that time it looked as if it would be so fantastically elaborate that it could never possibly be made. But the great developments in wireless and electronic valves during the war have altered the picture, because through complicated electric circuits you can do many things at enormously greater speed than you could do before, mechanically. Consequently, Turing, who is now on our staff, is showing us how to make his idea come true. Broadly we reckon that it will be possible to do arithmetic a hundred times as fast as a human computer, and this, of course, means that it will be practicable to do all sorts of calculations outside the scope of human beings. The machine will have many different parts, such as circuits which do addition or multiplication, or one might give them orders like this: 'Choose any number, then carry through with it a set of prescribed operations, and if the answer is bigger than seven go back and start again with a new trial number, but if it is less than seven, you are to do some other different operations.' A different part of the machine is its memory, for it works

[1] Quoted by kind permission of Sir Charles Darwin, F.R.S., and the British Broadcasting Corporation.

so fast that there is no time to write down the answers, and therefore, there must be gear which can remember many things, so as to have them handy when they are going to be needed again for later work. On the practical side of the problem a great deal of work has already been done in America, and a machine very much on these lines already exists there, though its rate is not as fast as we hope to get and its gear is a good deal more elaborate – that is often the penalty of starting first.

Needless to say the press quickly pounced on this and the caricaturists made great play with it. One evening paper went so far as to head a short paragraph about Alan with the words "Electronic Athlete." Indeed journalists' imaginations seemed to have been stimulated by reports of what computing machines might do. Some years later Alan remarked that the daily papers were many years ahead of him, opening even *his* eyes in wonder, so far did they outstrip him in their forecasts.

He considered that one of the best accounts of the future achievements of this Automatic Computing Engine, henceforward to be known as ACE, was to be found in *The Surrey Comet* in its issue of 9th November, 1946, from which I quote:

Some of the feats that will be able to be performed by Britain's new electronic brain, which is being developed at the N.P.L., Teddington, were described by Dr. A.M. Turing, thirty-four years old mathematics expert, who is the pioneer of the scheme in this country. The machine is to be an improvement on the American ENIAC, and it was in the brain of Dr. Turing that the more efficient model was developed. Dr. Turing, speaking about the 'memory' of the new brain . . . said, 'it will be able to retain for a week or more about as much as an actor has to learn in an average play.' Or to put it another way he said, 'the machine will be able quite easily to remember about ten pages from a novel, though not, of course, in their ordinary form. They would have to be translated into a medium it is capable of "understanding," in other words into the digits that it is designed to handle.'

As for the ability of ACE to play an average game of chess, Alan told the reporter that that was looking far into the future, and a more capacious and longer 'memory' would be required by ACE. Questioned about the power of judgment needed for playing chess and for other activities he conceded that this was a "matter for the philosopher rather than the scientist," adding, "that is a question we may be able to settle experimentally in about 100 years' time."

The Surrey Comet goes on to say:

The N.P.L. scientists are convinced that it will be the only one of its type that will ever be made. Improved versions will be on paper before it is ready. Instructions will be fed to the machine on packs of cards, on which certain patterns are punched. This it is expected will take about two minutes, compared with several hours on the American version, and ACE will work at extraordinarily high speeds. It will, for instance, multiply two ten figure numbers in two thousandths of a second. It will be expert also at simultaneous equations. The average mathematician has neither the time nor the patience to tackle one with more than a dozen unknowns; ACE will take one with a hundred unknowns, and solve it in its stride.

Leaders of the team developing ACE were Sir Charles Darwin, F.R.S. (Director of the N.P.L.), shortly afterwards succeeded by Sir Edward Bullard, F.R.S., Mr. J. R. Womersley (Superintendent of the Mathematics Division), and Professor D. Hartree, F.R.S., the only man in this country who had up till then worked the American ENIAC (Electronic Numerical Integrator and Computer). ACE was expected to have a higher memory storage than ENIAC.

Pending the fulfilment of Alan's project for a full-scale computing engine, a working model, known as the Pilot Ace, was built and an opening demonstration of its capabilities was given in the autumn of 1950, which was followed by a cocktail party given by Sir Edward and Lady Bullard to a number of scientists. The party over, Alan was invited by the host and hostess to an informal supper, during which he was much interested to hear something of the history of their official residence, Bushey Park, and of Queen Adelaide whose home it had been.

On 30th November, 1950, *The Times* published this account of the Pilot Ace:

The Ace itself will be built later, but the model demonstrated here today is none the less a complete electronic calculating machine, claimed as one of the fastest and most powerful computing devices in the world. Its function is to satisfy the ever-increasing need in science, industry, and administration, for rapid mathematical calculation which in the past, by traditional methods, would have been physically impossible or required more time than the problems justified. The speed at which this new engine works, said Dr. E. F. Bullard, F.R.S., Director of the laboratory, could perhaps be grasped from the fact that it could provide the correct answer in one minute to a problem that would occupy a mathematician for a month. In a quarter of an hour it can produce a calculation that by hand (if it were possible) would fill half a million sheets of foolscap paper.

The automatic computing engine uses pulses of electricity, generated at a rate of a million a second, to solve all calculations which resolve themselves into addition, subtraction, multiplication, and division; so that for practical purposes there is no limit to what ACE can do. On the machine the pulses are used to indicate the figure 1, while gaps represent the figure 0. All calculations are done with only these two digits in what is known as the binary scale. When a sum is put into the machine the numbers are first translated into the binary scale and coded; instructions are also given to the machine by coding them as holes in cards. To carry out long sequences of operations the engine must be endowed with a 'memory.' This 'memory' section is highly complicated. It depends upon the slower time of travel of supersonic waves, into which the electric pulses are converted, through a column of mercury. One thousand pulses – representing digits – can be stored in this way and extracted at the precise moment when they are needed by the 'arithmetic section,' which, handling pulses of electricity, is working 100,000 times faster than the supersonic section. The completed calculation appears in code as a holed card, representing the answer in the binary scale, which is translated back into ordinary numbers. When experience has been gained some improvements will doubtless be made to the Pilot Ace and embodied in the first standard prototype model. The cost of development and construction of the pilot model, which uses some 800 thermionic valves, was about £40,000. Now it is ready to 'do business' and is expected to more than earn its keep.

The accounts in the contemporary press are of historic interest, for a considerable impact was made on the public by the announcement of the construction and claims of the Automatic Calculating Engine, substantiated as the latter were by the actual performance of the Pilot Ace. Progress in certain respects, such as in playing chess, was accelerated more than Alan himself at first predicted.

In a letter to me Dr. E.T. Goodwin, now Superintendent of the Mathematical Division of the National Physical Laboratory, has traced the development there of computing machinery. He writes:

In the early years after the war Alan produced what we call the 'logical design' of a large computer which was to be called 'The ACE' or Automatic Computing Engine. The Laboratory was very doubtful of its ability to produce successfully what was then so ambitious a machine and, at about the time when Alan took his sabbatical year at Cambridge, it was decided to produce a small version which would be entitled the Pilot Ace. Though the basic ideas behind this machine were largely Alan's, you will understand that the detailed arrangement was decided by others. This machine did magnificent work for four to five years, during the latter part of which the English Electric Co. produced the DEUCE, the machine we now have, which is what we call an 'engineered version' of the Pilot Ace. This really means that it is a much more shiny and smart looking job, lacking the 'string and sealing wax' appearance of the typical lab. product; in essentials it is the same machine as the Pilot Ace. We have under construction the full-scale ACE. Again this embodies many of your son's ideas, but, of course, we have learnt a lot about the behaviour of these machines in the intervening years; new equipment has been developed, and so there are many ways in which it differs from his original plan. The original Pilot Ace was eventually given to the Science Museum to use as an exhibit, though unfortunately not a working one, which would have involved them in too much effort.

(This letter is dated February, 1957.)

The relegation of the Pilot Ace to the Science Museum brings home the rapid development of computing machinery in the last ten years, which recalls the passage on page 352 of *The Nature of the Physical World*, where the late Sir Arthur Eddington, F.R.S., graphically described the swiftly changing course of scientific investigation. He wrote[2]

Scientific Discovery is like the fitting together of the pieces of a great jig-saw puzzle; a revolution of science does not mean that the pieces already arranged and interlocked have to be dispersed; it means that in fitting on fresh pieces we have had to revise our impression of what the picture-puzzle is going to be like. One day you ask the scientist how he is getting on; he replies, 'Finely. I have very nearly finished this piece of blue sky.' Another day you ask how the sky is progressing and you are told, 'I have added a lot more, but it was sea, not sky; there's a boat floating on the top of it.' Perhaps next time it will have turned out to be a parasol upside down; but our friend is still enthusiastically delighted with the progress he is making.

While employed at the National Physical Laboratory, Alan lived at a guest house (Ivy House) in Hampton-on-Thames. As usual, he disregarded appearances; to help his hostess he used, in running kit, to climb the mulberry tree in the garden to pick the mulberries, getting covered with juice in the process. When there was a dearth of potatoes he discovered a shop where they were to be had and carried home a large sack of potatoes on his back. One way and another he won his hostess' admiration and respect.

During his time at Teddington he occasionally stayed near Dorking with Professor Champernowne's mother, who remembers his appearing in the drawing-room one evening with a pair of white socks which he proceeded to darn. The work finished, he said he had found it very soothing. Actually he darned extremely neatly and put a very neat darn in his dark-coloured trousers – but

---

[2] Permission to use this quotation has kindly been given by the Cambridge University Press and the Macmillan Company, New York (19th March, 1957), Copyright 1930.

unfortunately in white or some light colour. It was in Mrs. Champernowne's garden that he used to play a complicated form of chess with Professor Champernowne. In this game, each player had to run round the garden after his move, and if one arrived back at the board before his opponent had moved, he was allowed to have an extra move. Fleetness of foot probably helped to counterbalance Alan's not being a very good chess player. The purpose of this game was to throw light on the physiological effects of violent exercise on the functioning of the brain.

On many Sundays, having posted a change of clothes in advance, he would run the eighteen miles from Hampton to Guildford to see me. When, after the war, the Post Office was engaged in research on computers Alan was sometimes required to attend conferences at Dollis Hill and visit the Post Office laboratories. He disliked complicated cross-country journeys which involved underground trains, buses, etc. (and anyhow often left things behind in them), so he usually ran the fourteen miles from Teddington to Dollis Hill wearing some very old flannel trousers tied at the top with a piece of rope. (My protests about that rope were received with the information that it was a perfectly good bit of rope and did its job.) A colleague, travelling by train and bus, used to bring Alan's suitcase with some tidier clothes; they generally arrived almost simultaneously. During the rest of the day Alan drank large quantities of water. When one considers the whole proceeding from his point of view, one sees that he got the exercise of which he always made a great point, and avoided a wearisome journey; why, therefore, use the ordinary method to reach Dollis Hill?

This was not the only time that he was undeterred by distance. Once when he had gone to the Oxford and Cambridge Boat race and spent the day with friends, he arrived at Surbiton – just after the last train had left. Nothing daunted, he set out after midnight, and guided by the stars found his way to the main road and walked the sixteen miles home to Guildford, and then turned up punctually at breakfast, as if a long midnight walk were nothing unusual.

A Sabbatical Year at King's

In August 1947 my husband died. Alan, disappointed with what appeared to him the slow progress made with the construction of ACE, and convinced that he was wasting time since he was not permitted to go on the engineering side, asked for a sabbatical year. This was allowed, and he returned to King's, where he was still a Fellow, and was presumably occupied with research, for in 1948 *The Quarterly Journal of Mechanical and Applied Mathematics* published his paper "Rounding-off Errors in Matrix Processes," while "Practical Forms of Type-theory" appeared the same year in the *Journal of Symbolic Logic*. He doubtless enjoyed being once more in the academic atmosphere among old friends, and now he was able to indulge his new enthusiasm for psychology and physiology, of which more later.

About this visit and others to King's information has kindly come from Professor Pigou and Mr. Peter Matthews and Dr. J.W.S. Pringle. Professor Pigou mentions his coming in after Hall to play a game of chess and continues:

He was not a particularly good player over the board, but he had good visualizing powers, and on walks together he and an Oxford friend used to play games by simply naming the moves. This, from the point of view of a chess master, is very small beer – Alekhine once played thirty-three games simultaneously blindfold – but for us humble wood-pushers it was impressive. He told me he had deliberately cultivated this faculty by putting up in his bedroom pictures, at first of the separate quarters, and later of the whole board... Though primarily a mathematician, he was interested in many other things as well and would gallantly attend lectures on psychology and physiology at an age when most of us were no longer capable of sitting on a hard bench listening to someone else talking.

Possibly already he was interested in the processes involved in the form and growth of living tissues, and attendance at these lectures was in order to equip him to develop, as he did later, his original chemical theory of the growth of living things. He was then thirty-five and a don of twelve years standing, but in addition to the advanced lectures in physiology he occasionally went to the elementary ones, in the company of Mr. Peter Matthews, then in his second year and reading for the Natural Sciences Tripos. Mr. Matthews recalls "his refreshing outlook, as some of the things which puzzled the lecturers were clear to him, whereas many simple biological things he found puzzling as there was no physical explanation." In their discussions following the lectures he was fascinated by Alan's approach to the subject. He used sometimes to try to draw him out on his ideas, but Alan would very soon stop using words and break into mathematical symbolism, too difficult to follow. Mr. Matthews adds: "He introduced me to the similarities between computing engines and brains and I found and still find this a very useful comparison." He came to know Alan better when they spent a week together at Professor Pigou's lakeland cottage, Gatesgarth, where he found him a wonderful walking companion without, however, much head for heights when they ventured on to rock.

Their companionship was not limited to theory. Mr. Matthews wanted to determine the breaking strength of some rope to be used for climbing and tells how Alan "ingeniously suggested that we could determine the tension in it by its frequency of vibration, as we progressively loaded it by winding it on to the revolving spikes designed to prevent climbing into College. Unfortunately though we spent an enjoyable hour we didn't obtain a very satisfactory result." How illustrative this was of Alan's way of using the thing nearest to hand to support some theory.

In the sphere of mathematics Professor Pigou has this to relate.

Once I remember I put to him a matter in which I was having a discussion with an American economist, the full solution of which required some (to me) rather difficult mathematics. He found that I was wrong and the American right, but that the mathematical argument with which the American purported to support his case was wrong. He himself worked out what I presume was a valid argument, but would not let it be printed because he said, 'whatever it might be as economics, as mathematics it was not interesting.'

While away in Cambridge he wrote a report on "learning machines" for the National Physical Laboratory whither he returned about May 1948. As progress on the ACE had not come up to his expectations he sent in his resignation from the Scientific Civil Service. It was something of a shock to him to find himself summarily dismissed; quite probably he had completely forgotten the terms of his contract. Sir Ben Lockspeiser who was conversant with Alan's work at the N.P.L., since it came under the Department of Scientific and Industrial Research, expressed his opinion of him in these terms: "I was personally interested in his work for which I had a great admiration . . . His work remains. We were and are proud of him as a brilliant and lovable colleague and he will long remain in our memories."

Since this chapter was written the construction of ACE has been completed (1958) by the Control Mechanisms and Electronics Division of the National Physical Laboratory. Though further experience and better equipment have led to many divergences from the first design, Alan is still regarded as its originator: his, too, are the general logical ideas, especially the more revolutionary aspect of them. Consequently at the Press Day in November, 1958, Dr. A.M. Uttley, Superintendent of the above Division, declared, "Today, Turing's dream has come true": (thanks largely to Alan's friends' faith in his foresight). Not only has his dream come true, but his ideas have profoundly influenced computer designers throughout the world.

9

Work with the Manchester Automatic Digital Machine

On resignation from the Scientific Civil Service Alan accepted a Readership at Manchester University and was appointed Assistant Director of "Madam," the Manchester Automatic Digital Machine designed mainly by Professor F.C. Williams, F.R.S., and Dr. T. Kilburn. This machine at the time of its construction was reputed to have the largest memory storage capacity of any known machine and was expected to be able to "remember" a bookful of facts and figures.

As Assistant Director (but ignorant himself of the identity of the Director) Alan's role was to collaborate with Professor F.C. Williams and Dr. T. Kilburn and lead the mathematical side of work with the machine. According to Professor M.H.A. Newman's account in the Royal Society's Memoir, "for a few years he continued to work first on the design of the sub-routines out of which the larger programmes for such a machine are built, and then, as this kind of work became standardized, on more general problems of numerical analysis." With the experience so gained he produced in 1950, *The Programmers' Handbook for the Manchester Electronic Computer.*

In June 1949, a representative of *The Times* asked him by telephone sundry questions and put forward certain propositions regarding the scope of the machine; in reply Alan hazarded some forecasts of what the machine might eventually achieve. He was aghast a day or two later to find the results of this telephone conversation expanded into a long paragraph on the centre page of *The Times*.[1] In it he is quoted as saying:

We have to have some experience with the machine before we know its capabilities. It may take years before we settle down to the new possibilities but I do not see why it should not enter anyone of the fields normally covered by the human intellect, and eventually compete on equal terms. I do not think you can even draw the line about sonnets, though the comparison is perhaps a little bit unfair because a sonnet written by a machine will be better appreciated by another machine.

"Isn't that just like Alan?" commented a relative. *The Times* continued:

Mr. Turing added that the University was really interested in the investigation of the possibilities of machines for their own sake. Their research would be directed to finding the degree of intellectual activity of which a machine was capable, and to what extent it could think for itself.

At first on going north Alan lived in rooms at Hale, but later bought a house (Hollymeade, Adlington Road) in Wilmslow, Cheshire, and moved there about August 1950. He then wrote: "I think I shall be very happy here." He attacked with enthusiasm a mass of jungly growth at the end of his garden, and evidently got a poisoned scratch, for he developed acute sinovitis in his arm. He had no help, and, as he was picnicking with the barest necessities, it was very unfortunate. It was a relief to the doctor when he responded well to penicillin. He very much enjoyed his garden, but, though in *Who's Who* he entered gardening as one of his hobbies, his methods were crude and amateurish. Rather unexpectedly he insisted on the arrangement by himself of the flowers indoors, entirely in accordance with his own aesthetic ideas. What gave him great pleasure in his new home was having friends to stay with him or to dine. In the latter case the study was made unusually

---

[1] *The Times*, 11th June, 1949.

tidy and particular attention paid to the temperature of the wine and choice of the menu. He was a good host and, like his father, an entertaining conversationalist when in congenial company. The equipment of his house was curious. He tolerated an ill-assorted medley of shabby furniture and a few good pieces, but was almost faddy over the design and quality of his table-glass and china.

Mrs. Clayton, his invaluable housekeeper, and he shared many jokes, for he delighted to regale her with tales against himself. There was, for instance, the occasion when, his watch being under repair, he carried a little clock in his pocket. Suddenly in the crowded train to Manchester the alarm went off and everyone in the compartment jumped. On his runs he often forgot to take his door-key, so one was kept hidden near the spout of the garage gutter. One day he knocked it over the spout and it just slipped away into the ground, a fact which he reported to Mrs. Clayton with much relish.

He frequently rode the twelve miles to Manchester and back on an ordinary bicycle, to which he later had a motor attached. He paid little heed to being drenched with rain on these journeys, which were often taken in the middle of the night or in the small hours: it was not unusual for him to spend the whole night at work with the computing machine.

What much added to his contentment at Hollymeade was the great kindness of Mr. and Mrs. Roy Webb who occupied the semi-detached house adjoining his: besides this there was his fondness for their very small son, Rob, already mentioned. Mrs. Webb proved a most considerate neighbour in plying him with hot drinks when he had influenza; that is but one instance of her many acts of kindness and neighbourliness. On the days when his excellent housekeeper did not come he did his own cooking with some success and enjoyment, provided it did not claim too much time. Actually, I think he was a little proud of himself as a cook.

At Hale and Wilmslow he lived comparatively near Professor and Mrs. Newman, so that his friendship with them deepened and he was very much *ami de la maison*, walking in at the back-door whenever he liked. A workman, left alone in the house, reported a mysterious (but easily identified) visitor who had walked in by the back-door and out by the front door without a word of explanation. He often spent Sunday in Manchester with another family with which he was very much at home. Just three weeks before his death he passed with these friends a very entertaining day at Blackpool enjoying all the fun. Though Alan was then over forty their two young daughters (one only seven) were greatly attached to him. When the latter broke her right arm he brought her a lovely biscuit tin filled with sweets and chocolates; as he gave it to her he explained that it was a "left-handed" tin which she would be able to open herself. It was deemed the nicest present she had by reason of the thoughtfulness which prompted it.

At this period Mr. Denis Williams was once more in touch with Alan. Here in his own words is his last impression of him.

We had not seen each other for about a decade before he came to Manchester and in the meantime he had collected distinctions and a growing reputation, but these had made no difference to his manner. There was the same boyish grin, the same inability to take himself seriously (though his work he took very seriously indeed), the same modesty and personal austerity. The only side to his character which was new, to me at least, was his ability to play with children and apparently derive as much pleasure as he gave. My last memory of him is playing with our children in the snow, rolling snowballs to build an igloo. That was in the spring of 1954. I remember my wife saying then that she had never seen him look happier, and certainly the children thoroughly enjoyed themselves. Of his contribution to his own subject I am not able to judge, but it would be a dull mind indeed that did not recognize his extraordinarily keen intelligence; and to discuss any subject with him was a stimulating experience that made one aware of the confusion and sloth of one's own mental processes. In intellectual, as in other matters, it was essential to him that everything should ring true. Looked at in a different way, this may be responsible for the criticism that he could not accept the work of others. But it seems to me precisely this complete intellectual integrity, which, combined with his other gifts, made it reasonable to expect that he would produce results of fundamental importance in his own field.

Alan had a delightful sense of humour. He enjoyed elaborating fantastic projects, such as a scheme for faking prehistoric cave paintings, in mock-serious detail, or bringing an over-serious discussion down to earth with a quick colloquial turn of phrase. With him jest and earnestness were often closely intermingled.

I remember myself how he brought home a piece of flint from the Guildford Downs and spent a happy hour shaping it into an arrowhead which he despatched to an archaeological friend with the query, "What do you make of this?"

Cybernetics

Cybernetics[2] has been defined as the science of "Control and communication in the animal and the machine." Naturally experience with the Manchester Automatic Digital Machine encouraged Alan's expectations with regard to the possible scope of computing machines. Sometime round about 1944 he had talked to me about his plans for the construction of a universal computer and of the service such a machine might render to psychology in the study of the human brain. This he regarded as likely to be one of the more valuable contributions a universal computing machine could make to knowledge.

He must therefore have welcomed the opportunity to discuss all such matters with Professor Norbert Wiener. According to an article in *News Review* of 24th February, 1949, Professor Norbert Wiener, "a mathematical genius from Massachusetts University," had flown over to England, and besides seeing Professor D.R. Hartree, F.R.S., F.C. Bartlett and H. Levy, had long talks with Alan whom he acknowledged to be a pioneer, "first among those who

---

[2] The word "Cybernetics" was coined by Ampère, and is derived from the Greek for "steersman," from the same root as the word for "governor."

considered the logical problems of a machine as an intellectual experiment." The article appends a row of photographs of these learned men; in comic contrast is Alan's photograph which suggests its insertion by mistake, so young does he appear by comparison. Professor Wiener thought very highly of Alan's work which "represented an original combination of the modern stream of mathematical logic with the theory and practice of control and communication apparatus."

That there was a marked affinity of outlook between Professor Wiener and Alan may be seen from quotations here given from a paper by Abraham Kaplan, entitled, "Sociology learns the Language of Mathematics" (*The World of Mathematics*, Vol. II[3]) in which he describes Professor Wiener's conclusions as set out in his books, *Cybernetics* (Wiley, 1948) and *The Human use of Human Beings* (Houghton Mifflin, 1950). Abraham Kaplan's article runs:

> Cybernetics becomes relevant to the study of man because human behaviour is paralleled in many respects by the communication machines. This parallel is no mere metaphor, but consists in a similarity of structure between the machine processes and those of human behaviour. . . . Thus cybernetics bears on the study of human behaviour in a variety of ways: most directly, by way of neurology and physiological psychology; and by simple extension, to an improved understanding of functional mental disorders, which Wiener finds to be primarily diseases of memory, thus arriving at Freudian conclusions by a totally different route.

Drawing on Professor Wiener's book, *The Human Use of Human Beings*, he adds:

> 'it is now possible to arrange machines so that they can communicate with one another, and in no merely figurative sense. And in addition to electronic brains, machines can be equipped with sensory receptors and efferent channels.

---

[3] *The World of Mathematics*, "a small Library of the Literature of Mathematics from A'hmose, the Scribe, to Albert Einstein, presented with Commentaries and Notes by James R. Newman," published by Messrs. Simon and Schuster, New York, 1956.

The talks with Professor Wiener doubtless stimulated Alan's enthusiasm for comparing computing machines with the human brain, for his paper entitled "Computing Machinery and Intelligence," appeared in October 1950, in Vol. LIX, N.S., No. 236 of *Mind: a Quarterly Review of Psychology and Philosophy*. Almost the whole of this article is within the comprehension of the average reader and is highly entertaining. As Professor M.H.A. Newman in his Memoir of Alan, published by the Royal Society, wrote: "The conversational style allows the natural clarity of Turing's thought to prevail, and the paper is a masterpiece of clear and vivid exposition."[4] Not only is a brilliant case put forward for the theory of intelligent machinery, but it is of additional interest for the light thrown on Alan himself. His friends will recognize the characteristic touches, as it is rich in unexpectedly homely illustrations which forcibly and wittily drive home his points.

Under the title "Can a Machine Think?" it also appeared in Volume IV of *The World of Mathematics*, where the Editor describes Alan as "one of the most gifted modern mathematical logicians," and introducing the subject of computing machines writes:

Can machines think? Is the question itself, for that matter, more than a journalist's gambit? The English logician, A.M. Turing, regards it as a serious, meaningful question and one which can now be answered. He thinks that machines can think. He suggests that they can learn, that they can be built so as to be able to do more than we know how to order them to do, that they may eventually 'compete with men in all purely intellectual fields.' His conclusions are made plausible in the brilliantly argued essay below.

In a letter to me the Editor confessed himself privileged to be able to give it permanent form. Above this paper he has put three rather cryptic quotations:

"Thinking is very far from knowing." – Proverb.

---

[4] M.H.A. Newman (1955). Alan Mathison Turing. 1912–1954. *Biogr. Mems. Fell. R. Soc.* **1**, 246–252.

'Beware when the great God lets loose a thinker on this planet." – Emerson.

"For 'tis the sport to have the enginer hoist with his own petar . . . " – Shakespeare (*Hamlet*).

Mr. Denis Williams, already referred to as a contemporary at King's, regarded this paper as essentially *a jeu d'esprit*, and described it as "an outstanding contribution to a subject that has been much discussed in recent years," and adds, "he avoids the philosophical pitfalls with an adroitness that most professional philosophers might well envy. Once when asked how a computer could produce surprising results, his quick answer was that one got a bishop to talk to it." He generally had a ready reply to questions and criticisms. Asked under what circumstances he would say a machine was conscious he said that, if the machine was liable to punish him for saying otherwise, then he would say that it was conscious. With the development of his research on the resemblance between the human brain and universal computers he became involved in many discussions on the subject, when as *The Times* has it[5]: "his view expressed with great force and wit, was that it was for those who saw an unbridgeable gap between the two to say just where the difference lay." Apropos of these discussions Mrs. M.H.A. Newman notes:

> I remember sitting in our garden at Bowdon about 1949 while Alan and my husband discussed the machine ('Madam') and its future activities. I couldn't take part in the discussion and it was one of many that had passed over my head, but suddenly my ear picked up a remark which sent a shiver down my back. Alan said, reflectively, 'I suppose when it gets to that stage we shan't know how it does it.'

(Part II of this biography contains an account of Alan's theory of computing automata and a short paper by him on "Intelligent Machinery.")

---

[5] *The Times*, 16th June, 1954.

Ferranti, the firm which had built the Manchester Automatic Digital Machine, soon recognized Alan's qualifications as an authority on universal computers and in consequence appointed him one of their consultants.

In order to produce some likeness to the unpredictable element in human behaviour, he considered the introduction of a roulette wheel into a computing machine. The unexpected results would then enable machine owners to say with something akin to parental pride, "My machine (instead of 'my little boy') said such a funny thing this morning."

In his enjoyment of indoor games Alan wanted to get to the underlying principle; with this aim he began to write a paper on "The Theory of the Correct Strategy for Playing Two-man One-card Poker." His main interest, however, was in chess which, as he early foresaw, could be played by a machine. That he was not alone in holding this view is shown by Mr. Donald Michie's recollections, here recorded.

Alan told me that he and Champemowne had constructed a machine to play chess, in the sense of a complete specification on paper for such a machine. One could call it a 'paper machine' from which one could laboriously calculate move by move what the corresponding electronic machine would do were it constructed. Each move required perhaps half an hour's paper work as compared with the fraction of a second which a real machine would need. During a stay in Cambridge, Shaun Wylie and I constructed a rival 'paper machine' which we christened Machiavelli, from our two names, Michie–Wylie. On behalf of Machiavelli we then issued a challenge to the Turochamp (our name for the Turing–Champernowne machine), the game to be played by correspondence. Alan and I were responsible for conducting the correspondence and working out the moves for our respective machines. The labour involved proved too tedious for us and the game did not progress beyond the first few moves. Alan was then at Manchester, I think, and he had plans to programme the electronic computer there with the two chess machines so as to be able to run off a series of games between them in a short time and so discover which was the better. I think he embarked on this project, but it was never finished.

(Professor Champernowne tells me that the Turochamp 'paper machine' won against his wife who was a learner: it only took two or three minutes over a move; hence less than previous estimate.) As Assistant Director of the Manchester Automatic Digital Machine Alan was in a position to put to further test his belief in the ability of a machine to play chess. From the experience gained he contributed in 1953 an article, "Digital Computers applied to Games: Chess," to *Faster than Thought* (Editor, B.V. Bowden, Pitman, London.) Here he follows his preliminary remarks about machines and their design and programming to play chess, with the moves in an actual game between the machine and a moderate player. He then proceeds to criticize the machine's play and note its limitations. Elsewhere he expressed the view that a machine would not make the same mistake twice, unless the electric current were turned off, in which case it would be as fallible as any human player. Compared with playing chess, Solitaire would be simple for a machine to play and he thought a demonstration of its performance with Solitaire would be of interest to visitors and be easy to follow.

A not too abstruse article, "Solvable and Unsolvable Problems," came out in *Science News* (No. 31, Penguin Series, 1954). It begins with consideration of a puzzle consisting of a large square within which are smaller movable squares numbered one to fifteen and one empty space, into which any of the neighbouring movable squares can be slid, leaving a new empty space. "There are only a finite number of positions in which the numbered squares can be arranged (*viz.*, 20,922,789,888,000)." Note "*only.*"

In the course of the examination of solvable and unsolvable problems he makes some comparisons with knots – disentangling knots, changing one kind of knot into another without cutting the string. (My mind goes back to very early days when there were complaints of his spending the time fiddling with little bits of paper, etc.

Was he worrying out problems with knots, etc. ?) The article on "Solvable and Unsolvable Problems" ends characteristically with these words, "These, and some other results of mathematical logic, may be regarded as going some way towards a demonstration, within mathematics itself, of the inadequacy of 'reason' unsupported by common sense." Professor M.H.A. Newman's comment on this paper is that,

> it is essentially a popular account of the gist of the "Computable Numbers" paper. It was a remarkable feat to give a proof of the fundamental 'unsolvability' theorem understandable by readers with no previous knowledge but willing to think rather hard. The theorem is that there is no mechanical method (i.e., no automatic machine) for solving certain problems of which he gives a specimen. The fifteen puzzle and the knots are really just red herrings. (The first is given as an example of a problem which *is* soluble.)

Alan's remarkable but difficult paper, "The Word Problem in Semi-groups with Cancellation," was published in 1950 in *Annals of Mathematics* (Princeton). W.W. Boone has made a close study of this paper, and in his review, which appeared in the *Journal of Symbolic Logic*, Vol. 17 (1952), 742 has provided a useful discussion and analysis of it, and moreover cleared up innumerable misprints. The year 1953 saw the publication by the London Mathematical Society of Alan's paper "Some Calculations of the Riemann Zeta-function." Various incomplete drafts have been found among Alan's papers. These include, "The Reform of Mathematical Notation," containing no new ideas but intended as propaganda for mathematicians; and an unfinished paper entitled, "A Note on Normal Numbers," which is interesting as far as it goes.

Besides work on these mathematical papers, his research and routine duties at the University Alan found time for the supervision of a certain number of candidates for their Ph.D. degrees.

His election as a Fellow of the Royal Society took place in the spring of 1951; his proposer was Professor M.H.A. Newman, F.R.S., and seconder, Bertrand Russell, O.M., F.R.S. He thus became the fourth member of my family to be elected a Fellow of the Royal Society. He duly signed the historic Charter Book of the Royal Society, rather nervously and illegibly, as someone stood over him "guarding the book," as he reported, "with his life." He was very much gratified at his election. On the day it was announced Professor and Mrs. Newman invited him to supper to celebrate the event: he arrived with a bottle of wine in his bicycle basket. Following on this I gave a party for our friends in Guildford to meet the new F.R.S. I was a little surprised, in view of his retiring ways, that he readily consented to have this notice taken of his success.

In 1952 he resigned his fellowship at King's College, Cambridge, where he had been a Fellow for seventeen years. Next year he gave a lecture on computing machinery at his old school, Sherborne. Varied accounts reached me of this, especially of his delivery, for he walked up and down, probably trying to feel at ease lecturing in the presence of his former schoolmasters, but the subject was too intricate for this method, as well as being outside the capacity of most of his audience.

10

Broadcasts and Intelligent Machinery

On 15th May, 1951, on the Third Programme, Alan gave his first broadcast, which was one of a series with the general title "Automatic Calculating Machines." His lecture, the second in the series, bore the sub-title, "Can Digital Computers Think?" I heard at his house its recording, but he stoutly refused to join me, though he plucked up courage to listen to its repetition on 3rd July, 1951. It was, as far as I know, generally agreed that he put his case very clearly. The other participants in the series were: Professor M.H.A. Newman, F.R.S.; Professor F.C. Williams, F.R.S.; and Mr. M.V. Wilkes (later F.R.S.), who gave no sub-title to their lectures. These talks were all reasonably long, not those very short affairs that stop just as one is beginning to get interested.

These lectures were followed on 14th January, 1952, by a four-cornered broadcast discussion on the Third Programme between Professor Sir Geoffrey Jefferson, F.R.S., Professor M.H.A. Newman, F.R.S., Mr. R.B. Braithwaite, Fellow of King's College, Cambridge, and Alan. The subject was, "Can Automatic Calculating Machines be said to Think?" Alan vigorously maintained, with some mild support from Professor Newman, that they could be said to think and bore the brunt of their opponents' objections, particularly those of Sir Geoffrey Jefferson. The general impression of my non-expert friends was that the subject had been presented most amusingly and in such a way as to be, on the whole, within the comprehension of those to whom it was quite new. One recalls Sir Edward Appleton's words at a British Association meeting: "It is my belief that if a scientist cannot talk simply about his subject he has not got to the bottom of it himself." As he wound up the debate, Sir Geoffrey could not resist a final shot at Alan with these words, "It would be fun some day, Turing, to listen to a discussion, say, on the Fourth Programme between two machines on why human beings think that they think."

In his paper, "Computing Machinery and Intelligence," Alan had previously dissented from some of the views expressed by Professor Sir Geoffrey Jefferson in his Lister Oration, 1949, where he dealt with the scope of computing machines. Despite intellectual differences they were on the best of terms, as may be gathered from his letter of congratulation on Alan's election to a Fellowship of the Royal Society. It runs thus: "I am so glad; and I sincerely trust that all your valves are glowing with satisfaction, and signalling messages that seem to you to mean pleasure and pride! (but don't be deceived!)." Indeed, Sir Geoffrey had a deep admiration for Alan, as I know from a letter in which he spoke of "Alan in whom the lamp of genius burned so bright . . . he had real genius, it shone from him."

11

Morphogenesis

M orphogenesis is a term used to describe the processes involved in the appearance and development of living structures. How long Alan had been ruminating on his chemical theory of the growth of living things, particularly the mathematical aspect of it, we do not know. Whether in attendance at lectures on physiology in 1947 to 1948 he had this in view, or whether with his wide interests he studied physiology just to enlarge his knowledge and was thereby spurred on to further investigations, it is impossible to say. Whichever way it was, after going to Manchester, he began to develop his noted theory of morphogenesis which naturally led to the study of botany. On his walks and runs he was always on the look-out for flowers and grasses. With some advice from me and the help of *British Flora*,[1] (much worn after being carried about in his pocket), and *Flora of the British Isles*, by A.R. Clapham, T.G. Tutin and E.F. Warburg, he identified all those he found in his neighbourhood, and systematically marked on very large scale maps just where he had found the various specimens.

His lengthy monograph on morphogenesis was contributed in November 1951 to the *Philosophical Transactions of the Royal Society* under the title "The Chemical Basis of Morphogenesis"; this was revised in March 1952 and published by the Royal Society on 14th August, 1952, in the *Phil. Trans.*, Series B, No. 641. It was but a foretaste of what he hoped to develop more fully. With the

---

[1] *British Flora*, by George Bentham, C.M.G., F.R.S., revised by Sir J.D. Hooker, K.C.S.I., C.B., F.R.S. Seventh Edition revised by A.B. Rendle, F.R.S.

aid of the Manchester Computing Machine, he was able to carry out calculations connected with his new research. As Professor Newman wrote, in an appreciation contributed to the *Manchester Guardian* at the time of Alan's death:

In the last two years of his life he began to work out a remarkable chemical theory of the growth of living things. In this work he found the fullest scope for his mathematical powers, his great flair for machine computing, and his great power of tearing his way into a subject new to him – in this case the chemistry of living tissues.[2]

When he lectured on this subject great interest was aroused among biochemists; on one such occasion he was questioned for two whole hours. According to the King's College, Cambridge, Annual Report, November 1954:

he wanted to understand, at least in a simplified and idealized way, how it could happen that a spherically symmetrical mass of cells developed in the course of time into 'an animal, such as a horse, which is not spherically symmetrical'.[3]

In his mathematical approach to the study of form he used to devise, purely by mathematics, unsymmetrical curved shapes. He showed me some of these and asked whether they resembled the blotches of colour on cows, which indeed they did to such an extent that the sight of cows always calls to my mind his mathematical patterns.

In 1950 Professor J.Z. Young, F.R.S., delivered the Reith Lectures[4] entitled, "Doubt and Certainty in Science," being "a biologist's reflections on the brain." The subject was so closely allied to Alan's own work that I persuaded him to try, on his way through London,

---

[2] Copyright Guardian News & Media Ltd 1954.
[3] By kind permission of the Provost and Fellows, King's College, Cambridge.
[4] Published by the Clarendon Press, Oxford, 1951, with some additions and clarifications and reprinted in 1953 and 1956. To the work of both N. Wiener and Alan appreciative reference is made in the preface.

to see the Professor. A good deal of skirmishing by telephone resulted in a mutually profitable meeting; of this Professor Young has given an account, which he opened with words that have an amusing aptness.

My impression of your son is of his kindly teddy-bear quality as he tried to make understandable to others, ideas that were still only forming in his own mind. To me, as a non-mathematician, his exposition was often difficult to follow, accompanied as it was by funny little diagrams on the blackboard and frequently by generalizations, which seemed as if they were his attempt to press his ideas upon me. Also, of course, there was his rather frightening attention to everything one said. He would puzzle out its implications often for many hours or days afterwards. It made me wonder whether one was right to tell him anything at all because he took it all so seriously.

There seemed every hope that Professor Young and Alan might have complemented and advanced each other's research. The subject is more fully dealt with in Part II of this book in the chapter entitled "The Chemical Theory of Morphogenesis Considered."

Dr. J.W.S. Pringle, F.R.S., had introduced him to the Ratio Club. This was an informal group of friends associated with the Neurological Research Unit of the Medical Research Council, who used to meet for dinner in London to discuss things of mutual interest in biology and the physical sciences. On two occasions Alan submitted papers to be read. Dr. J.A.V. Bates, writing on behalf of the Ratio Club, said: "Alan used to come to our meetings when he could. Whenever he spoke he added something original, and he did it with much humour and authority. We have altogether the happiest memories of his visits."

For me to hazard a conjecture regarding one possible outcome of Alan's research in morphogenesis is truly a case of "fools rushing in where angels fear to tread," if ever there was one. However open to derision it may be, I put it forward. Since his theory of the chemistry of living tissues presumably had reference to normal tissues, it seems reasonable to hope that someone may be found to carry the work yet further and apply the theory to malignant growths and thus promote cancer research and open the way to the discovery of a cure.

12

Relaxation

Alan was a very hard worker, but when he relaxed he did so thoroughly, running, walking or cultivating his garden. On his short visits to Guildford in the vacations he always put in a good deal of work, yet he found leisure for walks with me and long discussions. Despite the great mental disparity between us we never lacked topics of conversation. I have never ceased to wonder at this ability on his part to share his interests and to step outside those realms of profound thought in which he lived. He interspersed with work a good deal of light-hearted fun. A sample of this was the Treasure Hunt at Leicester, when some of the clues were of Alan's invention. Thus he prepared, for each competitor, a bottle containing red liquid, either malodorous (labeled "The Libation") or drinkable ("The Potion"): when the bottle was emptied the next clue was revealed – written in red ink on the back of the label. As another clue he made up the word "*perplication*." Over his copy of *Les Faux Amis ou les Trahisons du Vocabulaire Anglais*[1] he put a convincing dust cover inscribed with the title, "Dictionary of Uncommon French Words." He then inserted the word "*perplication*" with an explanation in French involving references to Maimonides and treasure hunters. This done, he prevailed on a bookseller to place it on one of his shelves. Alan was particularly pleased with this clue. All quite absurd but it goes to show his enthusiastic enjoyment of such trifles: similarly, as Robin Gandy says, "He greatly and unaffectedly enjoyed devising and playing parlour

[1] By Maxime Koessler and Jules Derocquigny – Librairie Vuibert-Paris.

games; the more ingenious and original the rules, the more his pleasure."

For summer vacations Alan generally took a short holiday abroad. He and Neville Johnson went on a bicycling tour through France, over the Grand Massif in 1950, visiting the Lascaux Caves, which interested him very much. The following year he chose the French Alps and Switzerland. Very generous to others, he was austere in many of his habits, for he travelled third class abroad, put up at Youth Hostels, and in London made his way to the Y.M.C.A. rather than to an hotel. Sometimes he took a short holiday sailing at Itchenor with my husband or alone. Once on a trip alone to the Isle of Wight he narrowly escaped being run down by a large vessel. As a boy he enjoyed riding; so, what with rowing, mountain climbing, hockey, tennis, cycling and long distance running his range of exercise was varied, while for indoor games there was his interest in chess and "Go."

We met three or more times yearly: he came to Guildford, except in war-time, almost always for Christmas and at either Easter or Whitsun and again in the summer, as well as for an occasional night or two when he happened to be in London for a meeting. After he got his house at Wilmslow I visited him annually.

He often stayed with Professor and Mrs. Champernowne. Apparently once he considered that he knew them well enough to accept their invitation in an unconventional manner. This is Professor Champernowne's own account of the incident:

My wife and I had invited Alan to stay with us at Shotover around Christmas. One morning an envelope arrived containing a piece of perforated tape, and the postmark (Manchester) led me to suppose that this was Alan's eventual response to our invitation. Four hours of hard work broke the code and I learnt that he would arrive at 2 a.m. the next morning, and that a parcel of food which he was sending must be unpacked and immediately dealt with according to some specified instructions. My satisfaction in deciphering the message was damped the next day when Alan explained it had only taken him half a minute to type the message on to the tape, as it was in standard teleprinter code, and I gathered he had hardly supposed it would occupy more than a few minutes of my time to reverse the process.

As he had been to the trouble of sending a food parcel this may have been Alan's idea of a practical joke rather than a way of saving himself the effort of a letter.

In the summer of 1952 he broke quite new ground and travelled to Norway. This trip awoke his interest in Norwegian and the closely allied Danish language. On my last visit to him a year later, he was proficient enough to read to me some of Hans Andersen's less known fairy tales, translating direct from the Danish. For his summer holiday in 1953 he went, after a brief stay in Paris, to Corfu with the Club Méditerrané. He had hoped to mix more with the French members of the camp and exercise his French but found them, to his disappointment, rather aloof.

In boyhood Alan had been no reader of fiction as he much preferred an encyclopaedia or scientific work. In his late 'teens he read a certain amount of fiction but said he was hampered in his selection by reason of the nature of the titles. He had a particular fondness for *The Pickwick Papers*, George Borrow's books and Samuel Butler's *Erewhon*. This last possibly set him to think about the construction of an actual intelligent machine. He was always most faithful to the slow-moving classics such as *Vanity Fair*, Jane Austen's novels and Anthony Trollope's works: as well as reading them, he enjoyed the broadcasts of them in drama form. From modern books of fiction he chose Dorothy Sayers' novels and Galsworthy's *Forsyte Saga*. He often refused the offer of a book for railway journeys, as with pencil and paper he was content to pass the time doing mathematics. Poetry, with the exception of Shakespeare's, meant nothing to him: he liked people to state plainly what they had to say. The poet might reply: "What about your symbols?"

With great pleasure he read Tolstoy – Anna Karenina and then *War and Peace*, and found both books deeply interesting and satisfying. He wrote to the friend who had lent him the books of the sense he had of being a part of both novels, of finding himself and his friends in their pages. It was characteristic of his care and thoroughness that he made an immense family tree for *Anna Karenina* in order to keep all the connections in his mind as he read. (Actually genealogies interested him very much.)

He was a regular listener to the plays broadcast on Saturday and Monday nights, but above all he was a devotee of the broadcast "Children's Hour" programmes, and in common with younger listeners singled out the Toy Town programmes as his favourites. This was no affectation as has been suggested – affectation was quite foreign to his nature – but yet another demonstration of the childish streak which was part of his make-up. While listening he would continue some routine work that required only half his attention. Once when a particularly good fairy story was to be read or dramatised he telephoned to me from Wilmslow to recommend it to my notice.

Athletics

In the last ten years or so of his life Alan's principal form of exercise and relaxation was long distance running. At Teddington he went in seriously for it and joined the Walton Athletic Club. Very characteristically he was at pains, on joining the club, to give no hint of his background and abilities and Fellowship at King's. His strategy was wrecked, however, by the publication of news of the projected construction of ACE, in which Alan had such a share, and also when members of the Walton Athletic Club met members of the National Physical Laboratory in sporting events. His running was very successful, for he won the mile and three-miles race at the N.P.L. sports, and the three-miles club Championship in record time. He also won a three miles handicap at Motspur Park in 14 minutes 20 seconds with 20 yards to spare, his handicap being 360 yards. "That," as he wrote, "was the meeting at which all the stars were trying to break records, but in fact all pulling muscles instead. Being a humble athlete myself I was able to get away without pulling a muscle." Whatever his own estimate of his prowess, he was regarded by others as in the front rank. One who knew his form well considered that, given the time to train and apart from the injury received later, he would have been a great international athlete. C.J. Chataway and he were members of the Walton Athletic Club's team in the annual relay race from London to Brighton in 1950, but without success.

Just before Christmas 1946, he received orders to sail on the *Queen Elizabeth* to the U.S.A. on 26th December. He had a racing fixture for that day but, nothing daunted, he took a taxi from near Dorking where he had been spending Christmas, to Walton, and thoroughly enjoyed his three-miles race, an open event. He was running off scratch which, as he wrote, made him feel rather grand, but after a very exciting race he was beaten by a foot. He then had to speed back to Hampton to pack, and duly boarded the *Queen Elizabeth* at Southampton that evening. Friends had visions of him racing up the gangway in his running clothes at the very last minute. But it did not quite come to that. He very much enjoyed the good comradeship of the athletic club and sports meetings, and according to the Walton Athletic Club Secretary he was "a very good club man and greatly respected and liked by all the members there, as his election to the committee shows." The club comprised men from all walks of life – road-sweepers, clergymen's sons, dentists, clerks and so forth – he was always at ease among them and made them feel at ease. After an interval of some years they still talk about him. Even the professional photographer, whom I asked for permission to include in this book his photograph taken of Alan after a successful race, concluded his reply with these words: "May I close by saying that from what little I knew of your son,

I greatly admired him, as did all his colleagues." But the secretary reports Alan's absentmindedness in arriving for committees after lighting-up time with no lights on his bicycle.

He was Vice-President of the Walton Athletic Club from 1949 for five years and presented a cup known as the "Turing Cup" to be competed for annually in their 440 yards championship. The club was proud of his achievements, for he broke two records and in August 1947 came in fifth in the Amateur Athletic Association Marathon Championship at Loughborough College Stadium, Leicestershire, and this despite having to lie down on the roadside with cramp for what seemed to him an age. The distance was 26 miles 385 yards, and his time 2 hours 46 minutes 3 seconds, i.e., 13 minutes $43\frac{1}{3}$ seconds behind the winner. He probably reduced his form by travelling from Teddington to Loughborough on the day of the race, which he did to suit someone else's convenience – rather a pity as he was not a good traveller.

Encouraged by his success in the marathon he planned to enter for the forthcoming Olympic Trials and had many enthusiastic backers. Accordingly he did a good deal of training in long distance running while up again at King's, 1947 to 1948. I owe to Professor Pigou the following story:

Once in London a cross-country team finding itself a man short, asked him to make up the number. He had so far no idea that he could run [apart from some steeple-chasing at school] but came in, I believe, at the head of the field. Interested in this new accomplishment, he began to train seriously with a view to marathon racing and was to be seen with hair flying as he raced home at the end of the ten, fifteen or seventeen miles solitary 'scamper.' Most unfortunately he displaced something in his hip, and, in spite of the efforts of bone-setters, had to give up the idea of long distance racing. It would have been a feather in the cap of King's College to have had one of its Fellows coming in first in an Olympic marathon.

A member of the college recounted how on one occasion when he was out on a walk, Alan who was running, overtook him, but stopped to exchange a few words and then said, "I must push on as I am in a race and Wooderson is behind." Presently a group of runners came along led by Wooderson (a renowned athlete). When staying with his former housemaster at Sherborne Alan went out for a fifteen mile run, taking it "as casually," Mr. O'Hanlon said, "as we should take going to the post."

Casual as his training might seem to some, it was carried out according to a system. He kept a careful check on his weight and timing, and the method he applied to his running prevented his forcing his pace unduly. The story is told by Professor Champernowne's uncle that Alan set out from Westcott, near Dorking, to run up Leith Hill. But as his watch was presumably out of order, he ran with an alarm clock tied round his middle. It seems a little reminiscent of the crocodile in *Peter Pan*. One masseur whom he consulted about his injury was much amused because when he enquired of Alan when his leg hurt the reply was, "Oh! when I have run about twelve miles." The injury notwithstanding, he was one of a team of four from the Walton Athletic Club, which won a ten mile road race in May, 1950. (See the picture on 113.) But, though he continued up to the end to take long runs for exercise, the abstention from serious racing was a great disappointment: he missed the good fellowship of the competitors and the way in which they would cheer each other on – even when being surpassed themselves. While up at Cambridge, 1947 to 1948, he was invited to become a "second claim member" of the Achilles Club. The Walton Athletic Club, which he had previously joined had the first claim on him. It may seem strange in middle age to become a marathon runner, but it is recognized that men in their thirties and forties are more suited to long distance running than quite young athletes.

After a successful race. May, 1950

13

Last Days and Some Tributes

Alan's house had a room from which the bathroom had been
sliced, leaving a space – the "nightmare room" as he called
it – useless for domestic purposes. This he eventually turned into a
laboratory where he spent many happy hours carrying out experiments. According to some of his friends these were often of an
unnerving character. On one occasion Dr. Gandy refused to take
part in one such experiment in a laboratory alive with electrical
equipment, though he was prepared to pursue it in the kitchen.
It only goes to show Alan's readiness to take chances when possessed by some absorbing idea. One who knew him well wrote to
me that he was like a child when experimenting, not only taking
in the observed result mentally but testing it with his fingers. He
adds: "When we worked together on some electrical contraption
he several times got high-voltage shocks by sheer carelessness." One
experiment that he carried out to his own great satisfaction was
the gold-plating of the bowl of an egg-spoon: he used his grandfather's gold turnip-watch for the gold and, presumably, potassium
cyanide, hence the presence of the latter in solid form in a drawer
in his spare room.

Dr. Robin Gandy has given a full account of their joint activities over the week-end a few days before Alan's death. They were
busy with the preparation of a non-poisonous weed-killer and
sink-cleaner. "There was nothing sinister about these experiments,
which were just an example of Alan's liking to make things himself." He planned to make large quantities of sodium hypochlorite
(non-poisonous) for these two purposes: this would give him more

satisfaction than the purchase of ready-made cleaners and weed-killers. He always wanted to start from scratch and it was even said that he would have preferred, if possible, to make his own electric bulbs and batteries. He went so far as to contemplate making bricks to pave his garden path, but thought it came near to "cheating" to buy clay from a builder instead of digging it up himself. Some very rough unfired pottery in his laboratory suggests that he had made a start on work with clay.

On his last visit to Alan Dr. Gandy found a number of experiments in electrolysis in progress, all of the same kind.

They were [he says] of the 'desert island sort' – that is, not for discovering anything new, but for seeing how far one could go with entirely home-made apparatus. In one experiment coke was used as an electrode – and a very messy one, as after a bit it crumbled. I suggested using the carbon sticks out of old batteries, but Alan thought this would be breaking the rules of the 'desert island game.' Starting from, I think, common salt (and perhaps some other household substances which I have forgotten) he was seeing how many chemicals he could produce by electrolysis. All the solutions which were being electrolysed had weed juice added to them, made by crushing weeds and steeping them in water. Anyway the weed juice and coke combined to make all these experiments horribly murky and smelly. One of them almost certainly did produce potassium cyanide, and may have been intended to produce it: but this was not the aim of the experiments as a whole, which was to produce a wide range of chemicals from the simplest and most easily obtainable ingredients.

In Alan's laboratory was a spoon partially coated with some white substance and similar to the one he had previously gold-plated. Had he planned to plate another spoon, this time employing potassium cyanide of his own manufacture, it would have been thoroughly characteristic of his methods. Actually there was still in progress in his laboratory after his death, an experiment which smelt of bitter almonds. It was one of the "desert island" series, where coke was used as an electrode, and had been going on for three weeks.

Aged 35

On the 8th June, 1954, Alan's housekeeper found him dead in his bed; death was due to poisoning by potassium cyanide, and it presumably took place on the night of the 7th June. The verdict at the inquest was that the poison was self-administered, while the balance of his mind was disturbed. No poison was found in his bedroom. There was just a partly eaten apple on the table by his bed, for, as a rule, he used to eat an apple at night.

It is unlikely that Alan had any financial worries, since he had left a substantial credit at his bank. He was at the apex of his mental powers, with growing fame, and absorbed in his research on morphogenesis, which promised far-reaching results. By any ordinary standards he had everything to live for.

I can only narrate objectively details of the events noted by those in close touch with him on the days immediately prior to his death. On the 1st June he had entertained to dinner his next door neighbours, Mr. and Mrs. Roy Webb, when they passed, according to Mrs. Webb, "a most delightful evening." She saw him several times during the next two days; and on the 3rd June, the day Mr. and Mrs. Webb moved to Styal, Alan gave Mrs. Webb and her little boy, Rob, tea in the kitchen, where, joined later by Mrs. Clayton, they had a "jolly party." Alan was in the best of spirits and full of plans to visit the Webbs often on his way out from Manchester. He had been very glad to hear that their successors next door were young and had a small boy; so there was no expectation of loneliness. In the afternoon of the 7th June, the date of his death, he had met on his walk a neighbour and her child; as usual, he spoke to them and appeared not only his ordinary self, but very cheerful.

His housekeeper was away for the Whitsun Bank Holiday, the 7th June. On arrival next day she saw nothing more amiss than the light on in his bedroom. He had obviously cooked and eaten his late dinner, and, in his customary way, he had left the used cutlery and silver standing in a jug of water to be dealt with by her. On his writing-table ready for the post were acceptances of invitations for the near future, as well as tickets for the theatre, to which he was to take friends that very week. In his study were new socks, just bought. All, indeed, pointed to normal arrangements for the time ahead. Dr. Gandy wrote: "When I stayed with him the weekend before Whitsun, he seemed, if anything, happier than usual: we planned to write a joint paper and to meet in Cambridge in July."

Many friends, either by reason of his temperament and recent good spirits, or because of "his unlimited flow of ideas and great enthusiasm for putting them into practice," have been led to believe that his death was caused by some unaccountable misadventure. Besides, his inadvertence alone had always involved the risk of an accident.

There is a marked unanimity of opinion in the letters to me about Alan. Coupled with great admiration for "his profound originality and insight" there is repeated emphasis on his simplicity and integrity and complete "lack of pretentiousness and pomposity." Over and over again he is described as in a class by himself in the sphere of brain power. Mr. Denis Williams goes so far as to say, "his abilities were such that most of us could scarcely begin to appreciate them." However much others admired his brains and originality there was no question of his regarding himself as on a pedestal. Letter after letter particularly notes his lovable character and power to win affection, his humour and modesty. "M.H.A.N." in an appreciation[1] wrote:

Turing took a particular delight in problems, large or small, that enabled him to combine mathematical theory with experiments he could carry out, in whole or part, with his own hands. He was ready to tackle anything which combined these two interests. His comical but brilliant analogies with which he explained his ideas made him a delightful companion. It was, perhaps, a defect of his qualities that he found it hard to accept the work of others, preferring to work things out for himself. This undoubtedly slowed his work and made him a difficult author to read.

---

[1] *Manchester Guardian*, 11th June, 1954. Quoted by kind permission of the writer and the Editor of the *Manchester Guardian*. Copyright Guardian News & Media Ltd 1954.

Alan's flair for the bizarre but apt analogy also struck Dr. Gandy, who considered that the "mark of his genius was that even in the most abstract realms of thought he always bore in mind completely concrete cases and examples." He adds:

When we were engaged on war work, I always thought him a bit austere but at Cambridge I was enchanted to find how human he could be, discussing mutual friends, arranging a dinner-party, being a little vain of his clothes and his appearance. One of my happiest memories is of him and Nick Furbank and me playing a complicated game of hide-and-seek in the Botanical Gardens by moonlight.

Penetrating appraisal came from Dr. Milner-White, now Dean of York, who, as Dean of King's College Chapel had been in contact with Alan over several years. This is his tribute: "My affection for him was real . . . My respect for his mind and brilliance was beyond measure. And character? It was that of the true scholar, modest, generous, quiet – giving always the feeling that he saw behind many veils which we could not pierce."

Occasional visits to Sherborne kept warm the affection he had won as a schoolboy. One of his science masters, Mr. H.. Gervis, who must have taught many clever boys, expresses his memories of Alan thus:

Alan, the pupil I shall always remember best and be most proud to have known. So able, so modest, though I think really very assured, so much so that he never felt the need to show off, or to score off other people or parade his knowledge and ability. But to me, perhaps, the most lovable characteristic of his mind was the way in which he could be interested in any problem, however trivial. What fun we had once inventing electrical devices we knew would never be built, for counting bees as they flew in and out of their hives. I shall never forget showing him, years ago, an utterly erroneous answer I had from some pupil, just to remind him of the kind of material with which schoolmasters had to cope. He looked at it carefully and then pointed out that though it contradicted everything that the books had to say, yet there was nevertheless a grain of truth in what the boy had been trying to express; so humbling, yet so typical of the charitableness of his mind.

One last word from his housemaster, Mr. O'Hanlon, in his notice of Alan in the *Shirburnian*.

For those who knew him here, the memory is of an even-tempered, lovable character with an impish sense of humour and a modesty proof against all achievement. You would not take him for a wrangler, the youngest Fellow of King's and almost the youngest F.R.S. of his time, or a marathon runner, or that behind a negligé appearance he was intensely practical. . . . In all his preoccupation with logic, mathematics and science he never lost the common touch; in a short life he accomplished much, and to the roll of great names in the history of his particular studies added his own.

Alan's work shows throughout a unity of purpose which might not at first appear from the variety of titles borne by his published papers. This is stressed by Professor M.H.A. Newman in the Royal Society's Memoir[2] and thus described.

The central problem with which he [Alan] started and to which he constantly returned, is the extent and the limitations of mechanistic explanations of nature. All his work, except for three papers in pure mathematics (1935*b*, 1938*a* and *b*)[3] grew naturally out of the technical problems encountered in these inquiries. His way of tackling the problem was not by philosophical discussion of general principles, but by mathematical proof of certain limited results: in the first instance the impossibility of the too sanguine programme for the complete mechanization of mathematics, and in his final work, the possibility of, at any rate, a partial explanation of the phenomena of organic growth by the 'blind' operation of chemical laws.

---

[2] M.H.A. Newman (1955). Alan Mathison Turing. 1912–1954. *Biogr. Mems. Fell. R. Soc.* **1**, 246–252.

[3] 1935*b* Equivalence of left and right almost periodicity. 1938*a* Finite approximations to Lie groups. 1938*b* The extensions of a group.

A friend of mine, though neither a mathematician nor a scientist was, on reading the Royal Society's Memoir, greatly struck by this unity of purpose and compared Alan's ability to bring ideas into an ordered unity with that of the "masters of the spiritual life and saints who have made an ordered unity of the activities of the spirit." Such a comparison would surely have surprised Alan himself.

In the minds of those competent to judge there is no doubt that Alan's work will live. Not only was he a pioneer of universal computers in England, but he made a lasting contribution to the foundations of mathematics by his logical analysis of computing processes, while his research into morphogenesis remains classical.

As a permanent memorial, I have endowed the "Alan Turing Prize for Science" to be awarded annually at Sherborne School.

It is my hope that this short biography may serve to irradiate and perpetuate his memory. It is offered in gratitude for a richly endowed life, zealously devoted to the pursuit of knowledge and truth.

He does not sleep,

How could that eager mind be stilled by death?

(Evelyn Underhill)

PART TWO

Concerning Computing Machinery and Morphogenesis

14

Computing Machinery

The World of Mathematics,[1] referred to in Part I, contains an article entitled "The General and Logical Theory of Automata" by John von Neumann which gives the ensuing synopsis (vol. 4, pp. 2093–2095) of Alan's Theory of Computing Automata:

The English logician, Turing, about twelve years ago attacked the following problem. He wanted to give a general definition of what is meant by a computing automaton. The formal definition came out as follows:

An automaton is a 'black box' which will not be described in detail but is expected to have the following attributes. It possesses a finite number of states, which need to be *prima facie* characterised only by stating their number, say *n*, and by enumerating them accordingly: 1, 2, . . . , *n*. The essential operating characteristic of the automaton consists of describing how it is caused to change its state, that is, to go over from a state *i* into a state *j*. This change requires some interaction with the outside world, which will be standardized in the following manner. As far as the machine is concerned, let the whole outside world consist of a long paper tape. Let this tape be, say, one inch wide, and let it be sub-divided into fields (squares) one inch long. On each field of this strip we may, or may not, put a sign, say, a dot, and it is assumed that it is possible to erase as well as to write in such a dot. A field marked with a dot will be called a '1,' a field unmarked with a dot will be called a '0.' (We might permit more ways of marking, but Turing showed that this is irrelevant and does not lead to any essential gain in generality.) In describing the position of the tape relative to the automaton it is assumed that

[1] Published 1956. Simon and Schuster, New York.

one particular field of the tape is under direct inspection by the automaton, and that the automaton has the ability to move the tape forward and backward, say, by one field at a time. In specifying this, let the automaton be in the state i $(= 1, \ldots, n)$ and let it see on the tape an e $(= 0, 1)$. It will then go over into the state j $(= 0, 1, \ldots, n)$, move the tape by p fields $(p = 0, +1, -1; +1$ is a move forward, -1 is a move backward) and inscribe into the new field that it sees $f = 0, 1$; inscribing 0 means erasing; inscribing 1 means putting in a dot). Specifying j, p, f as functions of i, e is then the complete definition of the functioning of such an automaton.

Turing carried out a careful analysis of what mathematical processes can be effected by automata of this type. In this connection he proved various theorems concerning the classical 'decision problem' of logic, but I shall not go into these matters here. He did, however, also introduce and analyse the concept of a 'universal automaton,' and this is part of the subject that is relevant in the present context.

An infinite sequence of digits e $(= 0, 1)$ is one of the basic entities in mathematics. Viewed as a binary expansion, it is essentially equivalent to the concept of a real number. Turing, therefore, based his consideration on these sequences.

He investigated the question as to which automata were able to construct which sequences. That is, given a definite law for the formation of such a sequence, he enquired as to which automata can be used to form the sequence based on that law. The process of 'forming' a sequence is interpreted in this manner. An automaton is able to 'form' a certain sequence if it is possible to specify a finite length of tape, appropriately marked, so that, if this tape is fed to the automaton in question, the automaton will thereupon write the sequence on the remaining (infinite) free portion of tape. This process of writing the infinite sequence is, of course, an indefinitely continuing one. What is meant is that the automaton will keep running indefinitely and, given a sufficiently long time, will have inscribed any desired (but of course finite) part of the (infinite) sequence. The finite, premarked, piece of tape constitutes the 'instruction' of the automaton for this problem.

An automaton is 'universal' if any sequence that can be produced by any automaton at all can also be solved by this particular automaton.

It will, of course, require in general a different instruction for this purpose.

The Main Result of the Turing Theory. We might expect *a priori* that this is impossible. How can there be an automaton which is at least as effective as any conceivable automaton, including, for example, one of twice its size and complexity?

Turing, nevertheless, proved that this is possible. While his construction is rather involved, the underlying principle is nevertheless quite simple. Turing observed that a completely general description of any conceivable automaton can be (in the sense of the foregoing definition) given in a finite number of words. This description will contain certain empty passages – those referring to the functions mentioned earlier (j, p, f in terms of i, e), which specify the actual functioning of the automaton. When these empty passages are filled in, we deal with a specific automaton. As long as they are left empty, this schema represents the general definition of the general automaton. Now it becomes possible to describe an automaton which has the ability to interpret such a definition. In other words, which, when fed the functions that in the sense described above define a specific automaton, will thereupon function like the object described. The ability to do this is no more mysterious than the ability to read a dictionary and a grammar and to follow their instructions about the uses and principles of combinations of words. This automaton, which is constructed to read a description and to imitate the object described, is then the universal automaton in the sense of Turing. To make it duplicate any operation that any other automaton can perform, it suffices to furnish it with a description of the automaton in question, and, in addition, with the instructions which that device would have required for the operation under consideration.

As early as 1947 Alan was engaged in the study of the problem of the education of computing machines. In the following paper on "Intelligent Machinery" he examines this possibility. It is thought to be the script of a lecture given to a discussion group. It has therefore to be borne in mind that it was intended to be spoken and was probably not ready for publication.

Intelligent Machinery, A Heretical Theory

'You cannot make a machine to think for you.' This is a commonplace that is usually accepted without question. It will be the purpose of this paper to question it.

Most machinery developed for commercial purposes is intended to carry out some very specific job, and to carry it out with certainty and considerable speed. Very often it does the same series of operations over and over again without any variety. This fact about the actual machinery available is a powerful argument to many in favour of the slogan quoted above. To a mathematical logician this argument is not available, for it has been shown that there are machines theoretically possible which will do something very close to thinking. They will, for instance, test the validity of a formal proof in the system of *Principia Mathematica*, or even tell whether a formula of that system is provable or disprovable. In the case that the formula is neither provable nor disprovable such a machine certainly does not behave in a very satisfactory manner, for it continues to work indefinitely without producing any result at all, but this cannot be regarded as very different from the reaction of the mathematicians, who have for instance worked for hundreds of years on the question as to whether Fermat's last theorem is true or not. For the case of machines of this kind a more subtle kind of argument is necessary. By Gödel's famous theorem, or some similar argument, one can show that however the machine is constructed there are bound to be cases where the machine fails to give an answer, but a mathematician would be able to. On the other hand, the machine has certain advantages over the mathematician. Whatever it does can be relied upon, assuming no mechanical 'breakdown,' whereas the mathematician makes a certain proportion of mistakes. I believe that this danger of the mathematician making mistakes is an unavoidable corollary of his power of sometimes hitting upon an entirely new method. This seems to be confirmed by the well-known fact that the most reliable people will not usually hit upon really new methods.

My contention is that machines can be constructed which will simulate the behaviour of the human mind very closely. They will make mistakes at times, and at times they may make new and very interesting statements, and on the whole the output of them will be worth attention to the same sort of extent as the output of a human mind. The content of this statement lies in the greater frequency expected for the true statements, and it cannot, I think, be given an exact statement. It would not, for instance, be sufficient to say simply that the machine will make any true statement sooner or later, for an example of such a machine would be one which makes all possible statements sooner or later. We know how to construct these, and as they would (probably) produce true and false statements about equally frequently, their verdicts would be quite worthless. It would be the actual reaction of the machine to circumstances that would prove my contention, if indeed it can be proved at all.

Let us go rather more carefully into the nature of this 'proof.' It is clearly possible to produce a machine which would give a very good account of itself for any range of tests, if the machine were made sufficiently elaborate. However, this again would hardly be considered an adequate proof. Such a machine would give itself away by making the same sort of mistake over and over again, and being quite unable to correct itself, or to be corrected by argument from outside. If the machine were able in some way to 'learn by experience' it would be much more impressive. If this were the case there seems to be no real reason why one should not start from a comparatively simple machine, and, by subjecting it to a suitable range of 'experience' transform it into one which was more elaborate, and was able to deal with a far greater range of contingencies. This process could probably be hastened by a suitable selection of the experiences to which it was subjected. This might be called 'education.' But here we have to be careful. It would be quite easy to arrange the experiences in such a way that they automatically caused the structure of the machine to build up into a previously intended form, and this would obviously be a gross form of cheating, almost on a par with having a man inside the machine. Here again the criterion as to what would be considered reasonable in the way of 'education' cannot be put into mathematical terms, but I suggest that the following would be adequate in practice. Let us suppose that it is intended that the machine shall understand English, and that owing to its having no hands or feet, and not needing to eat, nor desiring to smoke, it will occupy its time mostly in playing games such as chess and 'Go,' and possibly bridge. The machine is provided with a typewriter keyboard on which any remarks to it are typed, and it also types out any remarks that it wishes to make. I suggest that the education of the machine should be entrusted to some highly competent schoolmaster who is interested in the project, but who is forbidden any detailed knowledge of the inner workings of the machine. The mechanic who has constructed the machine, however, is permitted to keep the machine in running order, and if he suspects that the machine has been operating incorrectly may put it back to one of its previous positions and ask the schoolmaster to repeat his lessons from that point on, but he may not take any part in the teaching. Since this procedure would only serve to test the *bona fides* of the mechanic, I need hardly say that it would not be adopted in the experimental stages. As I see it, this education process would in practice be an essential to the production of a reasonably intelligent machine within a reasonably short space of time. The human analogy alone suggests this.

I may now give some indication of the way in which such a machine might be expected to function. The machine would incorporate a memory. This does not need very much explanation. It would simply be a list of all the statements that had been made to it or by it, and all the moves it had made and the cards it had played in its games. These would be listed in chronological order. Besides this straightforward memory there would be a number of 'indexes of experiences.' To explain this idea I will suggest the form which one such index might possibly take. It might be an alphabetical index of the words that had been used, giving the 'times' at which they had been used, so that they could be looked up in the memory. Another such index might contain patterns of men on parts of a 'Go' board that had occurred. At comparatively late stages of education the memory might be extended to include important parts of the configuration of the machine at each moment, or in other words it would begin to remember what its thoughts had been. This would give rise to fruitful new forms of indexing. New forms of index might be introduced on account of special features observed in the indexes already used. The indexes would be used in this sort of way. Whenever a choice has to be made as to what to do next, features of the present situation are looked up in the indexes available, and the previous choice in the similar situations, and the outcome, good or bad, is discovered. The new choice is made accordingly. This raises a number of problems. If some of the indications are favourable and some are unfavourable what is one to do? The answer to this will probably differ from machine to machine and will also vary with its degree of education. At first probably some quite crude rule will suffice, e.g., to do whichever has the greatest number of votes in its favour. At a very late stage of education the whole question of procedure in such cases will probably have been investigated by the machine itself, by means of some kind of index, and this may result in some highly sophisticated, and, one hopes, highly satisfactory, form of rule. It seems probable however that the comparatively crude forms of rule will themselves be reasonably satisfactory, so that progress can on the whole be made in spite of the crudeness of the choice of rules. This seems to be verified by the fact that engineering problems are sometimes solved by the crudest rule of thumb procedure which only deals with the most superficial aspects of the problem, e.g., whether a function increases or decreases with one of its variables. Another problem raised by this picture of the way behaviour is determined is the idea of 'favourable outcome.' Without some such idea, corresponding to the 'pleasure principle' of the psychologists, it is very difficult to see how to proceed. Certainly it would be most natural to introduce some such thing into the machine. I suggest that there should be two keys which can be manipulated by the schoolmaster, and which can represent the ideas of pleasure and pain. At later stages in education the machine would recognize certain other conditions as desirable owing to their having been constantly associated in the past with pleasure, and likewise certain others as undesirable. Certain expressions of anger on the part of the schoolmaster might, for instance, be recognized as so ominous that they could never be overlooked, so that the schoolmaster would find that it became unnecessary to 'apply the cane' any more.

To make further suggestions along these lines would perhaps be unfruitful at this stage, as they are likely to consist of nothing more than an analysis of actual methods of education applied to human children. There is, however, one feature that I would like to suggest should be incorporated in the machines, and that is a 'random element.' Each machine should be supplied with a tape bearing a random series of figures, e.g., 0 and 1 in equal quantities, and this series of figures should be used in the choices made by the machine. This would result in the behaviour of the machine not being by any means completely determined by the experiences to which it was subjected, and would have some valuable uses when one was experimenting with it. By faking the choices made, one would be able to control the development of the machine to some extent. One might, for instance, insist on the choice made being a particular one at, say, ten particular places, and this would mean that about one machine in 1,024 or more would develop to as high a degree as the one which had been faked. This cannot very well be given an accurate statement because of the subjective nature of the idea of 'degree of development,' to say nothing of the fact that the machine that had been faked might have been also fortunate in its unfaked choices.

Let us now assume, for the sake of argument, that these machines are a genuine possibility, and look at the consequences of constructing them. To do so would of course meet with great opposition, unless we have advanced greatly in religious toleration from the days of Galileo. There would be great opposition from the intellectuals who were afraid of being put out of a job. It is probable though that the intellectuals would be mistaken about this. There would be plenty to do in trying, say, to keep one's intelligence up to the standard set by the machines, for it seems probable that once the machine thinking method had started, it would not take long to outstrip our feeble powers. There would be no question of the machines dying, and they would be able to converse with each other to sharpen their wits. At some stage therefore we should have to expect the machines to take control, in the way that is mentioned in Samuel Butler's *Erewhon*.

15

Theory of Morphogenesis Considered

D r. J.W.S. Pringle, F.R.S., of the Department of Zoology, Cambridge, in his paper on "The Origin of Life" which appeared in the *Symposia of the Society for Experimental Biology*, Number VII, Evolution, 1953, has, under the heading, "The Appearance of Heterogeneity," dealt with Alan's theory of morphogenesis and his mathematical approach to the subject. The relative passage runs;

Turing (1952) has demonstrated by mathematical methods that certain types of dynamic system which are initially homogeneous undergo a progressive change which leads to the appearance of spatial heterogeneity. The essential minimum for this behaviour to occur appears to be the presence of two substances, which he calls morphogens, the concentration of one of which is dynamically maintained by a balance between a generation process whose rate is controlled by its concentration (autocatalysis in the wide sense) and a destruction process; the second being formed at a rate dependent mainly on the concentration of the first and being destroyed at a rate proportional to its own concentration; and that there be movement from one point to another according to the laws of diffusion. Turing has so far published an account of the behaviour of only one set of equations showing this effect, namely, the linear differential equation for morphogens X and Y

$$\left.\begin{array}{l} \frac{dX}{dt} = 5X - 6Y + 1, \\ \frac{dY}{dt} = 6X - 7Y + 1. \end{array}\right\} \tag{5}$$

He has, however, (personal communication), also considered the more general problem involving non-linear terms. These equations are related to but not identical with those proposed by Volterra (1931), but the new feature of Turing's work arises from the simultaneous consideration of diffusion as a factor influencing the concentrations in a region of space. By variation of the constants and the form of the equations Turing finds cases in which stationary wave patterns of concentration are generated, and it is clear that where dynamic equilibria of the right type exist for more than one substance with interaction between the two chemical systems, a local concentration can occur without the intervention of adsorption phenomena on to pre-existing particles. This discovery opens up a new field of speculation and experiment, since it may be possible to devise conditions in which the concentration of free radicals becomes sufficiently high in a localized part of the reaction system to set in train a new reaction involving other substances and carry the synthesis of organic molecules to a more complex stage.

It follows from Turing's arguments that in a system which is initially completely homogeneous, the *position* at which local concentrations will appear is indeterminate, in the sense that it is 'caused' by chance fluctuations in the rates of the various reactions due to the fact that each molecular transformation is a discrete event. The system is unstable in respect of its local concentrations, and like any unstable system it may be started on its course towards stability by an event however small. The system therefore provides a means of creating structure where no structure was initially present. If, however, there is some initial heterogeneity due to other factors, this can provide the initial stimulus for morphogenesis if the heterogeneity has a component of its structure similar to the inherent tendency of the system. Under these circumstances the instability of the dynamic system acts as an amplifier of certain preferred qualities of the initial state, building them up to a macroscopic scale and suppressing other qualities which do not fit in with the tendencies. An analogous situation is presented by any self-oscillatory system on the verge of instability which will amplify disturbances near the frequency of its inherent oscillation, and if no external disturbances are presented will select that frequency component from the thermal noise energy fluctuation in its parts, reaching its limiting amplitude of oscillation in any case, but with the *phase* of oscillation in this case indeterminate.

Among Alan's papers were found rough notes of his later research in his chemical theory of morphogenesis. Dr. N.E. Hoskin and

Dr. B. Richards have studied these and the conclusions they have drawn from what they have been able to piece together are to be published under the title "The Chemical Theory of Morphogenesis, Part II, and the Morphogen Theory of Phyllotaxis". A short summary of their earlier findings appeared in the Royal Society's Memoir,[1] (pp. 261–262) and follows here.

Chemical Theory of Morphogenesis

The work falls into two parts. In the first part, published (1952) in his lifetime, he set out to show that the phenomena of morphogenesis (growth and form of living things) could be explained by consideration of a system of chemical substances whose concentrations varied only by means of chemical reactions, and by diffusion through the containing medium. If these substances are considered as form-producers (or 'morphogens' as Turing called them) they may be adequate to determine the formation and growth of an organism, if they result in localized accumulations of form-producing substances. According to Turing the laws of physical chemistry are sufficient to account for many of the facts of morphogenesis (a view similar to that expressed by D'Arcy Thompson in *Growth and Form*.)

Turing arrived at differential equations of the form $\frac{\partial X_i}{\partial t} = f_i(X_1, X_2, \ldots, X_n) = \mu \nabla^2 X_i$, $(i = 1, 2, \ldots, n)$ for n different morphogens in continuous tissue; where f_i is the reaction function giving the rate of growth of X_i and $\nabla^2 X_i$ is the rate of diffusion of X_i. He also considered the corresponding equations for a set of discrete cells. The function f_i involves the concentrations, and in his 1952 paper Turing considered the X_i's as variations from a homogeneous equilibrium. If, then, there are only small departures from equilibrium, it is permissible to linearize the f_i's, and so linearize the differential equations. In this way he was able to arrive at the conditions governing the onset of instability. Assuming initially a state of homogeneous equilibrium disturbed by random disturbances at $t = 0$, he discussed the various forms instability could take, on a continuous ring of tissue. Of the forms discussed the most important was that which eventually reached a pattern of stationary waves. The botanical

---

[1] M.H.A. Newman (1955). Alan Mathison Turing. 1912–1954. *Biogr. Mems. Fell. R. Soc.* 1, 246–252. Quoted by kind permission of the Royal Society, Professor M.H.A. Newman, F.R.S., author of the Memoir, and Dr. N.E. Hoskin and Dr. B. Richards.

situation corresponding to this would be an accumulation of the relevant morphogen in several evenly distributed regions around the ring, and would result in the main growth taking place at these points. (The examples cited are the tentacles of *Hydra*, and whorled leaves.) He also tested the theory by obtaining numerical solutions of the equations, using the electronic computer at Manchester. In the numerical example, in which two morphogens were supposed to be present in a ring of twenty cells, he found that a three or four lobed pattern would result. In other examples he found two-dimensional patterns, suggestive of dappling; and a system on a sphere gave results indicative of gastrulation. He also suggested that stationary waves in two dimensions could account for the phenomena of phyllotaxis.

Dr. Hoskin has now extended the above with the following summary of the results of his and Dr. B. Richard's further research. I am greatly indebted to them both for their expert assistance.

This work was continued by considering quadratic terms in the reaction functions in order to take account of larger departure from a state of homogeneous equilibrium. This was still being developed at the time of his death and unfortunately much of the material is in a form which makes it extremely difficult to discover the results he obtained. This is particularly true of the numerical computations he was carrying out on the Manchester University Electronic Computer. However enough can be sifted to show that consideration of the quadratic terms in the reaction rates is sufficient to determine practical solutions in certain simple cases, whereas linear terms are really only sufficient to discuss the onset of instability.

The equations he considered were a development on those given above viz:

$$\frac{\partial U_j}{\partial t} = \left[-\phi(-\nabla^2)U\right]_j + G^{(1)} V_j^2 + 2G^{(2)} V_j U_j + G^{(3)} U_j^2,$$

$$\frac{\partial V_j}{\partial t} = \left[-\phi(-\nabla^2)V\right]_j + F^{(1)} V_j^2 + 2F^{(2)} V_j U_j + F^{(3)} U_j^2,$$

where U_j is the concentration of the jth morphogen and V_j is the concentration of the jth growth-retarder or "poison" in the medium. The function $\phi(-\nabla^2)$ is a certain linear operator, the characteristic vectors of which are the same as those of the operator "∇^2" but where $-\nabla^2$ has the characteristic value α, $\phi(-\nabla^2)$ has the characteristic value $\phi(\alpha)$. Also $\phi(-\nabla^2)$ has a maximum near $-\nabla^2 = K_0^2$ so that only components with wavelengths near to $2\pi/K_0$ will be significant. Certain assumptions are made concerning the reaction rates viz. $F^{(1)} = F^{(2)} = G^{(1)} = 0$, and it is also assumed that the poisons V_i are in effective equilibrium i.e.,

$$\frac{\partial V_j}{\partial t} = 0.$$

For small organisms he considers that the functions V_i are independent of position i.e., the organism is so small that the poison is assumed to be uniformly diffused through it. In addition $U_j(t)$ must be a linear combination of eigenfunctions with the same eigenvalue, i.e., there will only be one wavelength so that the operator $\phi(-\nabla^2)$ becomes $\phi(\alpha_0) =$ constant. Then if $\mathcal{F}$ is a linear operator which removes all components except those with the appropriate wave-length it is shown that the equilibrium solutions are solutions of the equation

$$U = \mathcal{F}(U^2).$$

For a sphere the operator $\mathcal{F}$ is one which removes from a function on a sphere all spherical harmonics except those of a particular degree. Thus to solve the above equation is to find a spherical harmonic of that degree which, when it is squared and again has the appropriate orders removed, is unchanged. For each degree there is only a finite number of essentially different solutions, i.e., solutions which are inequivalent under rotations of the sphere. These were investigated by Dr. B. Richards and results were obtained which could be compared extremely satisfactorily with the biological species *Radiolaria*. These marine organisms are unicellular and are surrounded by a skeleton, generally composed of silica, for support and protection. This skeleton is spherical, about a millimetre in diameter and has radial spikes which radiate from the outer shell of the skeleton. The number, arrangement and disposition of the spikes is usually the determining factor regarding the general form of the skeleton. The life of a single cell is essentially individual and its growth is affected by its surroundings. Thus it is reasonable to think that the various forms which abound are due to various concentrations of diffusing organisms both organic and inorganic (e.g., the salinity of the water of the silica content of the shell) and determine whether a similar distribution of local concentrations can be obtained from the equation

$$U = \mathcal{F}(U^2).$$

Dr. Richards found that by considering spherical harmonics of various degrees such corresponding solutions could be found. For example, considering the harmonics of degree six a solution is

$$U = A\overline{P_6^0}(\cos\theta) + B\overline{P_6^5}(\cos\theta)\cos 5\phi,$$

$\overline{P_6^m}(\cos\theta)$ being the normalized Legendre associated function (see Hobson, E. W., *Spherical and Ellipsoidal Harmonics*, Cambridge, 1951). This solution corresponds to a regular icosahedron with twelve local maxima being situated regularly around the sphere. This configuration is reproduced in the subclass *Phaeodaria* of *Radiolaria*, some of which have twelve equal and equidistant radial spines on a spherical ground form, the bases of the spines being situated at the apices of a regular inscribed icosahedron.

Other forms can be found (e.g., there is a solution of degree four which has six local maxima with a corresponding physical specimen in the legion *Spumellaria*) by varying the degree or finding other distinct solutions for the same degree, and in the majority of cases it is possible to find a corresponding skeleton from among the species of *Radiolaria*. While not conclusive this certainly seems strong evidence that the theory given by Turing is based on sound physical argument and that more complicated growth systems might be reproduced if the theory were developed further.

Geometrical Theory of Phyllotaxis

In this mathematical discussion of the geometry of phyllotaxis (i.e., of mature botanical structures) he considered ways of classifying phyllotactic patterns and suggested various parameters by which such patterns may be described. The subject of phyllotaxis deals with the arrangement of leaves on the stems of plants, and by a liberal interpretation of the terms "leaf" and "stem" it deals also with the arrangements of florets in a head (e.g., in a sunflower) and with the leaf primordia in a growing bud. The first part of the discussion was concerned with mature structures and considered leaves as geometrical points on a cylinder. He discussed various systems of co-ordinates which could be used to describe the phyllotactic lattice of points, many of these systems, of course, being in common use by descriptive botanists. As an example, if one considers a branch of *Pinus*, the scales which formed the base of the leaves lie at remarkably regular intervals on the branch. A line joining successive scales forms a helix of constant pitch and successive leaves subtend approximately constant angles at the centre. Such a system is defined by Turing (and some botanists) by means of three parameters, *viz*:

(a) *the jugacy J*, where *J* is the number of leaves lying at one level on the stem,
(b) the *plastochrone distance η*, corresponding to the portion of the pitch of the helix needed to move from one leaf point to the next, and
(c) *the divergence angle α*, the corresponding angle of rotation about the axis. Given these three parameters and the position of a single leaf it is possible to reconstruct the entire lattice.

This consideration of mature specimens was undertaken as a preliminary to his theory of morphogenesis which is concerned with the concentration of chemical substances. These concentrations are responsible for the form taken by the leaf primordia and they in turn determine the shape of the mature lattice of leaf positions. This description of phyllotactic patterns should therefore not be read as a botanical essay but rather in the light of a prologue to his completely mathematical theory of morphogenesis. Thus some of his methods of describing phyllotactic lattices would probably not be considered practicable by botanists although they are mathematically precise. The most fundamental way of describing a lattice is by a matrix,

$$\begin{pmatrix} a & b \\ c & d \end{pmatrix}$$

where (a, b) and (c, d) are vectors generating the lattice, i.e., each point is of the form $M(a, b) + N(c, d)$. This has the advantage of generality but its main disadvantage is its lack of uniqueness. One may make it unique by using specific co-ordinates such as the co-ordinates, J, η and α defined above together with an extra co-ordinate to fix the scale. Turing uses ρ, the radius of the cylinder and the matrix description becomes

$$\begin{pmatrix} 2\pi\rho\tau^{-1} & 0 \\ \alpha\rho & \eta \end{pmatrix}.$$

Such a description is sufficient to determine the lattice because every point in the lattice is of the form $(M2\pi\rho\tau^{-1} + N\alpha\rho, N\eta)$. However, such a description is suitable only for mature specimens and for considering lattices nearer the growth apex where the elements of the matrix may be considered to be continuously changing with respect to some independent variable e.g., time. Turing proposed the use of 'flow matrices.' If $A(t_0)$ is some matrix description of the lattice at some time to then the 'flow matrix' of the process is defined as $[A(t)]^{-1}A'(t)$, the dash representing differentiation with respect to t. This flow matrix is independent of the matrix description chosen, for if $B(t)$ is another matrix description there is an improper unimodular matrix L such that $B(t) = LA(t)$ and if $A(t)$ and $B(t)$ are continuous L must be constant. But then

$$[B(t)]^{-1}B'(t) = [LA(t)]^{-1}LA'(t) = [A(t)]^{-1}A'(t).$$

If one again considers helical co-ordinates so that the matrix description is

$$\begin{pmatrix} 2\pi\rho\tau^{-1} & 0 \\ \alpha\rho & \eta \end{pmatrix},$$

while the flow matrix of the process is

$$\begin{pmatrix} F_{11} & F_{12} \\ F_{21} & F_{22} \end{pmatrix} = \begin{pmatrix} \frac{d\log\rho}{dt} & 0 \\ \left(\frac{\rho}{\eta}\frac{d\alpha}{dt}\right) & \frac{d\log\eta}{dt} \end{pmatrix}.$$

A convenient way of picturing flow matrices is to imagine the change in the lattice as being due to the leaves being carried over the surface of the lattice by a fluid whose velocity is a linear function of position. The flow matrix then gives the relation between the velocity and the position. This point of view is particularly suitable when one is concerned with leaves which are sufficiently mature that they no longer move with respect to the surrounding tissue but only have movement due to the growth of that tissue. The coefficient F_{11} then represents the exponential growth in girth of the stem and the coefficient F_{22} the exponential rate of increase of the stem in length. The sum of these, the trace of the flow matrix, is the exponential rate of increase of the leaf area. The coefficient F_{21} represents whatever tendency there is for the stem to twist. It should be small, or in other words the divergence angle should not be appreciably affected by such growth. If this coefficient F_{21} is zero the flow may be described as being 'without twist.' A flow without twist and with $F_{11} = F_{12}$, i.e., a scalar flow matrix, may be described as a 'compression'; while a flow with $F_{11} + F_{22} = 0$ may be described as 'area preserving.'

He considers other ways of describing lattices but much of these descriptions are obviously part of a mathematical attack carried out to find the most practical method of description for his purpose. These will be published in the hope that botanists may find some of them useful but it is obvious that Turing did not intend this theory to be complete in itself but as a preliminary attack in any attempt to understand the mechanism operating at the growth apex.

An article by *The Times* science correspondent on "Recent Speculations on the Origin of Life," which appeared on 11th January, 1957, refers to Alan's approach as a mathematician to the subject. As this article puts it, "He [Alan] showed theoretically that, if the rates of formation and destruction of two substances were related in a defined way, then local concentrations of them occur spontaneously."

A letter from Alan to Professor J.Z. Young, F.R.S., with details of his research on brain structure and his mathematical theory of embryology is included here, together with Professor Young's comments.

Hollymeade,
Adlington Road,
Wilmslow.
8th February, 1951.

Dear Young,
I think very likely our disagreements are mainly about the uses of words. I was of course fully aware that the brain would not have to do comparisons of an object under examination with everything from teapots to clouds,

and that the identification would be broken up into stages, but if this method is carried very far I should not be inclined to describe the resulting process as one of 'matching.'

Your problem about storage capacity achievable by means of N (say 1010) neurons with M (say 100) outlets and facilitation is capable of solution which is quite as accurate as the problem requires. If I understand it right, the idea is that by different trainings certain of the paths could be made effective and the others ineffective. How much information could be stored in the brain in this way? The answer is simply MN binary digits, for there are MN paths each capable of two states. If you allowed each path to have eight states (whatever that might mean) you would get 3 (sic.) MN. If you want to get the sort of number McCulloch talks about you have to assume that e.g., 'dendrite combinations' can become facilitated, e.g., that when impulses arrive simultaneously on dendrites 3, 15, 47 the neuron is stimulated and for various other combinations, but this is not due to individual facilitation of these synapses. Moreover the particular combinations which will be facilitated will not be fixed at birth but determined by training. You could likewise have axon combinations or even a mixed kind. Since the number of subsets of 100 is 2^{100}, this sort of thing, if one were inclined to believe in it, would permit storage capacities of the order $10^{10}2^{100}$, the number of states being $2^{10^{10}2^{100}}$.

I am afraid I am very far from the stage where I feel inclined to start asking any anatomical questions. According to my notions of how to set about it, that will not occur until quite a late stage when I have a fairly definite theory about how things are done.

At present I am not working on the problem at all, but on my mathematical theory of embryology, which I think I described to you at one time. This is yielding to treatment, and it will so far as I can see, give satisfactory explanations of–

  (i) Gastrulation.
 (ii) Polyogonally symmetrical structures, e.g., starfish, flowers.
(iii) Leaf arrangement, in particular the way the Fibonacci series (0, 1, 1, 2, 3, 5, 8, 13, . . .) comes to be involved.
(iv) Colour patterns on animals, e.g., stripes, spots and dappling.
 (v) Patterns on nearly spherical structures such as some Radiolaria, but this is more difficult and doubtful.

I am really doing this now because it is yielding more easily to treatment. I think it is not altogether unconnected with the other problem. The brain structure has to be one which can be achieved by the genetical embryological mechanism, and I hope that this theory that I am now working on may make clearer what restrictions this really implies. What you tell me about growth of neurons under stimulation is very interesting in this connection. It suggests means by which the neurons might be made to grow so as to form a particular circuit, rather than to reach a particular place.

Yours sincerely,

A. M. Turing.

Extract from Professor Young's reply, 3rd March, 1951.

Thank you very much for your letter some time ago, which helped me a lot in thinking about these problems. I find that I take them rather slowly. I am very glad to hear that you are beginning to work on the embryological problems. I entirely agree that they are likely to yield to treatment, and they are not unconnected with the other problem. In fact it is most satisfactory to find someone who realizes this. I do hope you will be able to keep me informed of how things are going.

Of Alan's letter, Professor Young writes to me:

I have only one letter of your son's, which deals mainly with the question of whether one can make an estimate of how much information can be stored in a brain with a given number of nerve cells. The questions that I had asked him had to do with the problem of what anatomical investigations of brain structure would be likely to be useful. His letter attempts to solve this question, but unfortunately the information available on the subject is still very slight. As you will see in his last paragraph, your son emphasizes the connection between analyses of this sort, even when they do not seem to be related. I have always felt that his insight into the significance of morphogenesis was one of the best signs that he had really understood the biological problems. So many mathematicians who attack these problems seek only to understand the adult organism. Every biologist knows that a proper understanding can only come by studying how the tissues were formed and how they were maintained. Your son seemed to grasp this better than any other mathematician that I know . . .

So many mathematicians are concerned only with the relation between symbols, but your son was always prepared to be interested in the use of symbols in describing real things . . . I know too little to be able to be quoted as a really good judge of his character . . . All I could do was to admire from afar the results that he produced.

Professor Young thus saw very great possibilities in Alan's approach to morphogenesis, as another letter to me later shows. "Such contact as we biologists had with him [Alan] was a very stimulating experience . . . His techniques were rather above my head, but I always felt sure that it was just this sort of contact between the mathematician and the biologist that would ultimately make our science exact. The contributions that he published in this direction will certainly be classical."

Dr. Pringle holds that the implications of Alan's research on morphogenesis have not yet been fully appreciated. Years may elapse before a true estimate can be formed of his work in this sphere. As the late Archbishop William Temple stated in a quite different context: "Every great man is greater than his followers at first appreciate; it is posterity by which he is truly understood."[2]

---

[2] William Temple, *Readings in St. John's Gospel*, 1939, Macmillan & Co., Limited, pages xxv and xxvi. Reproduced with permission of Palgrave Macmillan.

MY BROTHER ALAN

≈

M y mother has written a biography of my brother Alan. It was certainly a *tour de force* on her part to write it, as she did, in her seventies. It has rightly been praised, by others better qualified to judge than I am, as a revealing study of a son who became, undoubtedly, a mathematical genius.

I start off, therefore, at a disadvantage. Much of the ground has already been covered, and certainly, so far as Alan's childhood is concerned, far better and in more detail than I could have done it myself. But it is sufficiently obvious to any discerning reader of that book that a false note has been struck somewhere: could Alan have been quite that paragon of virtue that my mother describes? Yet, if an elder brother ventures to suggest the contrary, it can easily be suggested that he is jealous and sour. This is a risk I must accept. My only concern is to put the record straight, however hazardous the enterprise.

My brother Alan was born on 21 June 1912 in a London nursing home. For me, it was a halcyon time, for my father, perforce, had to look after me for the one and only time in his life. His solution of the problem could not have been bettered: we visited the White City, went on roundabouts, ate in restaurants and travelled around the metropolis on the tops of buses with the tickets stuck in our hat-bands. Perhaps he overdid it a bit, for I was not accustomed to such undivided attention. So it did not seem to me at all a bad thing that my mother should be taking a nice long "rest" – a convenient euphemism for her confinement. I was not a little astonished and put out when I was taken to the nursing home one day and found that I had a new baby brother – Alan Mathison – whose arrival portended termination of the "rest." Was my nose put out of joint? It certainly was.

At this, and at all other times, my father took all decisions of consequence in the family. Now, rightly or wrongly, he decided that he and my mother should return alone to India, leaving both children with foster parents in England. Probably it was the right decision for me, for I had given my parents a bad fright with my dysentery in India and by the time my father was due for long leave again, I should be seven and a half. But it was a harsh decision for my mother to have to leave both her children in England, one of them still an infant in arms. This was the beginning of the long sequence of separation from our parents, so painful to all of us and most of all to my mother.

I am no child psychologist but I am assured that it is a bad thing for an infant in arms to be uprooted and put in a strange environment. I cannot speak for Alan but it was certainly a shock for me, even at the age of five. Not that it was my parents' fault; it was the accepted procedure for those who served the British Empire in India and elsewhere to entrust their children to foster parents in England. Who shall blame them? Even so, both of us were, in our different ways, sacrificed to the British Empire. I wish I could discuss the subject with Alan now, for he would surely have some very original views on it; for my part, being the minor sufferer, I think it was in a good cause, for I decline to subscribe to the current cant on the subject. Rudyard Kipling was no stranger to the subject and dealt with it adequately. My brother and I were lucky to escape the rigours of the life depicted by Kipling and were, indeed, fortunate in the home which my mother, with great diligence, found for us. But the ache remained. Moreover, the unsettled existence of our childhood was to leave its mark on us both.

Alan and I were left with "the Wards" – always we referred to them as "the Wards." We were the wards and they were our guardians but no matter – this was to be the centre of our existence for many years and our home from home. There we remained, on and off, for about eight years, except when our parents came home on leave from India at intervals of about three years. In many ways both of us felt more at home there than we did when our parents were on leave and we were living in a rented house in Scotland or in lodgings.

I believe it was here, perhaps in the first four or five years at the Wards, perhaps even in the first two, that Alan became destined for a homosexual. Has anyone mentioned it until now? No. My mother was fully aware of it before Alan's death (not, I imagine, that she had the faintest idea of what it implied), but she makes no reference to it in her book. One can put that down to Edwardian reticence if one pleases. In my view, based on such conversation as I had with my mother about it, necessarily reduced to a minimum, her reaction was much what one might expect if a specialist had informed her that her son was colour blind or had an incurable obsession with spiders: it was a nasty shock of brief duration and of no great significance. I am trying to make this memoir as truthful as I can, so I will not go to the length of pretending that I like homosexuals. To my mind, what is intolerable is the world of the "gay crusade" and, as my unfortunate brother may be cast in the part of an early and valiant crusader, this is by no means an irrelevant comment.

But to return to the Wards. The Wards lived at the top of Maze Hill at St Leonards-on-Sea. The name of the house was Baston Lodge and it stood (perhaps still stands) on a corner site opposite St John's Church, flanked on the downward gradient by Sir Rider Haggard's house, with its archway across the road. When I was older, I used to sit on the garden wall devouring *King Solomon's Mines*, *Allan Quatermain*, *She* and *Jess*, hoping to see the great man, for he was reputed to dictate these masterpieces in the arch-way room, but never did I catch a glimpse of him. My brother, more fortunate, once actually penetrated into the establishment for, mooning along the gutter of Maze Hill one day (he always preferred the gutter to the pavement), he picked up a diamond and sapphire ring, the property of Lady Haggard. He was sent round to the front door with it and was rewarded with thanks and a florin.

The head of the household was Colonel Ward – spare, gruff and taciturn, with eyes of the palest blue. His military bearing and manner concealed a warm heart, but of this I knew nothing in those early days and when it was, at long last, revealed to me, the habit of self-induced terror was paramount and I could not respond. There was also Mrs Ward, their four daughters – Nerina, Hazel, Kay and Joan – an assortment of Mrs Ward's powerful Haig nieces (she was vaguely related to Field-Marshal Haig, to say nothing of the Duke of Wellington, sub nom Wellesley) and stray young boarders such as Alan and myself. Mrs Ward became known to us as "Granny Ward" and, despite her military connections and obvious sympathies and a rock-like determination on Alan's part and mine to have nothing to do with either, she became very dear to both of us.

Let it not be supposed that Alan was a misogynist. The basic question was whether the female of the species was "safe" or not. Granny quite clearly was; so also was Hazel, who must have been of a very advanced age, at least twenty years old, in those early days, and, in any case, she had been in love with the much-married Mr King, the Rector of St John's across the road, for as long as anyone could remember. The fact that it was a hopeless quest did not seem to worry Alan.

The truth is that Hazel was no beauty, but, as Nanny used to say (they all said it), beauty is only skin deep. Hazel had to make do with being a saint and a saint she certainly was. Many years later, after her mother died, she achieved her life's ambition and became a missionary. It is very typical of my brother, who grudged every penny spent on himself, that he generously, and of his own accord, financed Hazel in this. Quite unknown to me, he had kept in touch with her for thirty years or more, and they had remained firm friends.

Kay, the third daughter, was anything but "safe"; she had no sooner "put her hair up" (circa 1916, aged about 18) than she married an RFC pilot and disappeared to the Argentine. The eldest, Nerina, was, in my humble opinion, very safe indeed, but a first cousin of hers, confusingly also called Colonel Ward, was evidently not of that persuasion and several years later, she married him and she in turn departed.

That left the youngest, whom I have described elsewhere as a premenopausal afterthought. She was halfway in age between Alan and myself. I regarded her as a little pest. My brother rated her a tyrant. One thing is certain and that is that she was a thoroughly spoilt little brat and extremely self-willed into the bargain. It was not her fault; that was the way she had been brought up. But she was emphatically not "safe." Once, on St Leonard's beach, driven over the brink of insensate fury by Joan Ward, I hurled a well-aimed pebble and cut her lip open. For this I was rightly chastised by Granny Ward. What drove me to this extremity I have no idea, but I do know that, in my view, justice had been done all round.

Again, it is not for me to appraise the Joan Ward syndrome so far as Alan was concerned. Let the psychiatrists make of it what they will; no doubt they could have a field day of it. I just think, myself, that if Alan had been blessed with, let us say, an understanding and loving sister a year or two older, or even younger, than himself, instead of a sardonic elder brother and the tiresome Joan Ward, he might have turned out differently.

Some years ago, I compared notes with my cousin, Hyacinth, the eldest of the three cousins who later came to spend their holidays at the Wards. She was some six years older than myself and much better qualified to form an objective view of Joan Ward in those early days. Her view, surprisingly, partook more of Alan's prejudicial outlook than mine. She reminded me of how Joan Ward returned home after a visit to friends and was observed with disgust at the tea table. "Aren't you pleased to see Joanie?" asked one of the doting sisters. "No" replied Alan, sullenly. "I prefer the cat", added my cousin Ralph.

My mother, perhaps unwittingly, gives the impression in her book that she recognized Alan's genius from the start, and that she sedulously fostered it. If so, she did not give that impression in the family at the time; in fact, quite the contrary.

My father, on the whole, either ignored my brother's eccentricities, or viewed them with amused tolerance but (as will appear) there were deep dudgeons when Alan started to accumulate appalling school reports at Sherborne.

As for myself, with the selfishness of youth, and separated by a gap of four years, I did not care what Alan did, and I was content to go my own way, as indeed he was content to go his. Our interests were so dissimilar that they never clashed. (Nonetheless, I take credit for persuading my parents to send him to Sherborne instead of Marlborough, which all but crushed me and would certainly have crushed him. My parents also deserve credit for accepting my advice.)

The only person in the household who was forever exasperated with Alan, constantly nagging him about his dirty habits, his slovenliness, his clothes and his offhand manners (and much else, most of it with good reason) was my mother. If this was due to some early recognition of his genius, she was certainly doing nothing to foster it by trying to press him into a conventional mould. Needless to say, she achieved nothing by it except a dogged determination on Alan's part to remain as unconventional as possible. I concede that my mother was determined that something should be made of Alan and that she showed much perseverance in this exercise, but that is another matter.

More discerning than any of these was the headmaster of the prep school to which we boys were sent. If he had been a little more discerning, or if Alan had been less obstinate in his refusal to admit anything to his scheme of things except mathematics, geography and the art of origami, he might even have groomed

Alan for the stardom of a scholarship at Sherborne, but in those days it must have seemed a preposterous idea. I have some published evidence of the headmaster's discernment in school songs which he composed in doggerel and from which the following is an extract:

"Little Turing's fond of the football field
For geometric problems the touch lines yield."

Others were not so much to the point, for example:

"Little baby brother makes paper balls in droves" [origami]

And another reading:

"And no map will make an elder brother take a lower place."

This last was a wholly unnecessary reference to the shameful episode of Alan (bottom of the lowest form) beating me (top of the top form) in a free-for-all geography exam.

At this prep school, there was a splendid master by the name of Blenkins. I suspect that he was not too bright and, being able to sympathize with slow learners, he was an excellent teacher. He could even explain the mystic symbol x to the dimmest. But his explanation infuriated Alan. "Absolute rubbish ...," said Alan, "... he doesn't know what he is talking about." He would then explain the true nature of x at great length. For my part, I preferred Blenkins' version.

At the time when Alan was due to start his first term at Westcott House, Sherborne, my parents were living in Dinard so that my father could escape the ruinous British income tax, then, I think, at the rate of 4/6 (say 23p) in the pound. Consequently, we boys used to commute to and from school via Southampton and St Malo on the Channel ferry. On the day that Alan arrived at Southampton to start his first term, the General Strike of 1926 had started. He was quite equal to the occasion. He sent a telegram to his prospective housemaster, Mr O'Hanlon, to say that he would be arriving on the following day. He then sped off on his bicycle and put up at the Crown Hotel, Blandford, for the night, where he was charged 5/6 plus 6d tip for dinner, bed and breakfast. (My

mother says six shillings, but I feel sure he left a 6d tip). Even
in those days, six shillings (say 30p) could hardly be described as
exorbitant.

In later years, when Alan's bad reports, slovenly habits and
unconventional behaviour had tried Mr O'Hanlon to the utter-
most, he would remind himself of Alan's arrival during the General
Strike by way of consolation. But at the end of the Michaelmas term
1927, even the mild and understanding Mr O'Hanlon allowed his
exasperation to spill over into Alan's reports:

> "No doubt he is very aggravating; and he should know by now that I
> don't care to find him boiling heaven knows what witches' brew by the aid
> of two guttering candles on a naked window sill. However he has behaved
> very cheerfully and undoubtedly has taken some trouble, eg with physical
> training. *I am far from hopeless.*" [Italics supplied.]

Comments of this kind – and far worse from some form masters –
made no impression whatever on Alan. My mother, however, was
constrained to suppress every report until my father had been
fortified by breakfast and a couple of pipes. Alan would then be
given a lecture in my father's study. His only recorded comments
were "Daddy should see some of the other boys' reports," and
"Daddy expects school reports to read like after dinner speeches."
Personally I found it a good time to be out of the house.

The truth of the matter, as I now view it in retrospect, is that
neither of Alan's parents or his brother had the faintest idea that this
tiresome, eccentric and obstinate small boy was a budding genius.
The business burst upon us soon after he went to Sherborne. After
a few terms, it became apparent that he was far ahead of the other
boys in mathematics: when Alan was sixteen, the maths master
told my mother that there was nothing more that he could teach
him and he would have to progress from there on his own.

I think it must have been when Alan was due to take the School
Certificate examination (now replaced by "O" levels) that he read
Hamlet in the holidays. My father was delighted when Alan placed

the volume on the floor and remarked "Well, there's *one* line I like in this play." My father could already see a burgeoning interest in English literature. But his hopes were dashed when Alan replied that he was referring to the final stage direction (*Exeunt, bearing off the bodies*).

Alan was first class at beating the system. He refused to work at anything except his precious maths and science, but he had an incredible aptitude for examinations, aided by last minute swotting. At Sherborne, marks were awarded both for term's work and examination results; these were read out in turn, followed by the combined result. On one famous occasion, he was twenty-second out of twenty-three on the term's work, first in exams and third on the combined results. This made much sport for the Philistines but it did not endear him to the Common Room. For my part, viewing these exploits from afar, I thought it was great.

The question whether or not Alan should be allowed to sit for the School Certificate examination was an occasion for strife in the Common Room. Those on the mathematical and science side, convinced they were backing a winner, agitated for Alan to take the School Certificate; those teaching classics, languages, English literature, *et al.*, deemed it a shocking waste of time even to allow him to try. By this time the headmaster was one C.L.F. Boughey, who had tried (unsuccessfully) to teach me some Latin at Marlborough. Mr Bensly, who had a special form which he called "the Vermisorium"[1] for School Certificate candidates felt so passionately on the subject that he offered to give a billion pounds to charity if Alan should so much as *pass* in Latin. Boughey, nevertheless, ruled that Alan should sit the exam. Alan then proceeded to apply his mind to the subjects on offer and obtained credits in seven subjects, including Latin, English and French. There is no evidence that Mr Bensly paid up the billion pounds.[2]

[1] "The Wormery."
[2] Worth considerably more then than now!

Rumours of these matters reaching me, I began to realise that my brother was becoming a power in the land. Since nursery days, I had often pondered the story of the goose which turned into a swan. In the Brown, Yellow, Green, Blue and Red Fairy books, it was an unwritten rule that the younger (usually, one must concede, the third) son should make good; so also it was the third daughter, Cinderella, who captured the affection of the Prince. There was, indeed, in one of those many Fairy Books one story where the writer, bored, besotted or drunk, had made the *eldest* the hero, but despite being the first born myself, I deplored this departure from tradition. Now, suddenly, all was coming true as in the Fairy Books. Alan was making good. My father and I suffered successive phases of disbelief, scepticism and recognition as Alan's scholastic achievements smote us in rapid succession after the manner of Samson's jaw-bone of an ass.

My mother evidently persuaded herself that Alan's sudden and unexpected blossoming at Sherborne, which had dumbfounded my father and myself, was no more than what she had anticipated all along. It would be more accurate to say that she had a blind faith in Alan, which, as everyone knows is the ability to believe in what you know to be untrue, and indeed she deserves credit for it. For my part, I was much shaken to learn that, after all these years of bawling Alan out, putting him into his complicated sailor suits, ensuring his punctuality at meals and enduring him as a bolshy newcomer to my prep school, he had suddenly made good. I was very proud of him, too: he was a proper job of a swan.

Alan's eccentricities at Sherborne were not confined to scientific experiments at Westcott House. Thus one January he swam the Yeo for a bet and one June he went on an OTC parade wearing an overcoat and no tunic. Some of these stories I heard from an Old Shirburnian when I was abroad in the war. I have no reason to disbelieve them. The Yeo swim was unfortunate because it was the precedent for another silly bet many years afterwards when Alan dived into a lake in January, contracted fibrositis and thereby put himself out of the Wembley Olympics marathon.

Nor were his eccentricities confined to Sherborne. One Easter holiday in Dinard, he spent all his time collecting seaweed and brewing it up in the cellar until at length he extracted a few drops of iodine which he carried back to the science master at Sherborne in high triumph. When later we were living in Guildford, he had a series of crazes. He tried to learn the violin, which was excruciating. Then he turned his attention to breeding those little red banana flies in test tubes, so that he could prove Mendel's theory at first hand. Unfortunately, they escaped and the house was full of banana flies for several days. Oddest of all, in the heat of summer, he spent much of his time dressed as a private soldier allegedly drilling at Knightsbridge barracks, to what purpose nobody knew, but looking back on it now, I strongly suspect that drilling was not the object of the exercise at all. He was, as I have said, good at beating the system and, of course, the odder the things he did, the less one was likely to enquire into them.

In 1931, Alan went to King's College, Cambridge, with a mathematical scholarship. Soon afterwards, in March 1935, he wrote the thesis for which he was awarded a Fellowship, at the unusually early age of twenty-two. I have a horrid recollection of "The Gaussian Error Function" (whatever that might be), reputedly the subject of this monograph, for Alan had left it to the eleventh hour to sort the sheets, parcel and dispatch them. My mother and I spent a frantic half hour on hands and knees putting them in order; Mother did up the parcel in record time and Alan sped with it to the GPO[3] on his bike, announcing on his return that there were at least twenty minutes to spare. This is my only positive contribution to mathematical thought.

On or before the outbreak of war, Alan was recruited from King's College, Cambridge, along with other promising mathematicians,

---

[3] The Post Office.

to work as a code-breaker at Bletchley. Much has been written about
this establishment in recent years, including a great deal of rubbish
about Alan; one author, who shall remain nameless, describing
him as the son of his maternal grandfather. Of course, none of us
knew then, nor for many years after the war, what Alan was doing
at Bletchley, where he spent most of the early part of the war. In
fact, he was engaged in breaking the German naval codes. The best
and shortest book on the work at Bletchley is Peter Calvocoressi's
Top Secret Ultra, published by Cassell, and this is what he wrote
about Alan:

> The breaking of the cyphers used by the German army in North Africa
> and the breaking of the U-boat code in December 1942 were achievements
> of singular significance. The latter, associated above all with the name of
> Alan Turing whom even BP's most brilliant cryptographers put in a class
> of his own, is exceptional both for its strategic consequences and for
> its technical complexity (the German navy having introduced the extra
> wheels which made the cryptographer's task the more formidable).[4]

Mr Herbert Marchant, reviewing the second volume of *British
Intelligence in the Second World War*, said:

> When, however, miraculously, Bletchley Park began to break the
> immensely complicated U-boat Enigma code, near-defeat turned dra-
> matically into a resounding victory.

As someone rightly remarked, for thus saving the nation from
disaster, Alan should have been given an earldom but, in fact, he
was awarded an OBE, which, to the amusement of his friends, but
quite properly, in my opinion, he kept in a tin box along with such
aids to gracious living as screws, nails, nuts and bolts.

Mr G.M. Watkins, in a letter published in *The Times*, 12 September 1981, suggested that someone should write a book about the
war-time eccentrics of Bletchley Park. I warmly support such a
suggestion. In the top league of eccentrics with my brother were

---

[4] *Top Secret Ultra* is currently available in a revised edition, published in 2011 by M. & M.
Baldwin.

such characters as the absent-minded Dillwyn Knox, brother of Ronald Knox, and "Josh" Cooper, of whom it was said that he had been seen walking out of the gates of Bletchley with his hat in his hand and his briefcase on his head.

My brother's eccentricities have become legendary and some of them have become distorted in the process. For example, Mr Ronald Lewin, in his book, *Ultra Goes To War*, says that "... in some fit of despondency Alan converted all his money into cash and buried it in the Bletchley woods as a reserve against disaster." In fact, he did nothing of the kind. He had decided that if there were a German invasion, banking accounts would be useless, so he bought some silver ingots for use on the black market. These he trundled in an ancient perambulator and buried in a field (not at Bletchley), where he made a sketch map of their position so he could find them after the war. After the war, he enlisted the help of his friend, Donald Michie (now Professor Michie of Edinburgh University), to dig up these ingots – using, typically, a homemade metal detector – but the heavy ingots were by now well on their way to Australia and were never seen again.

I hope the author of the book recommended by Mr Watkins will be sufficiently acute to distinguish eccentricity from idiosyncracy and both from the just plumb crazy. I rather think that my brother ran the whole gamut. In the "just plumb crazy" class, I put the business of his chaining his mug to the radiator to prevent its being stolen. This may, however, have been one of his practical jokes because he was heard to declare that he had devised a special code for the lock which he defied the other cryptographers to decipher.

At Bletchley (where Alan was known as "the Prof."), he used to cycle to and from work and in the summer, he would wear his civilian gas mask to ward off hayfever. This apparition caused consternation to others on the road. Some would search the skies for enemy aircraft and others would don their gas masks just to be on the safe side. It was again at Bletchley that for some reason Alan was attached to an army unit for a week, where he was treated with that brand of tolerant amusement which the armed forces reserve for boffins. This did not suit Alan at all, so when he heard that there was to be a cross-country race, he asked modestly if he could join in. The request was granted; all in the mess looked forward eagerly to the Prof. trailing in well behind the rest. Alan had failed to mention that he was a marathon runner of near-Olympic standard (soon after the war, in August 1947, he came fifth in the Amateur Athletic Association Marathon Championship). So, of course, he arrived back three minutes before anyone else. As practical jokes go, I think it was fair enough. Further details of his athletic exploits can be read in my mother's book. I cannot forbear to remark that the whole business was yet another unpredictable eccentricity, for he had never shown the slightest interest in athletics before, yet here he was, at the age of thirty-five, in the front rank of marathon runners and seriously considered for the Wembley Olympics.

My mother gives a true picture of Alan's generosity. I have already mentioned Hazel, who achieved her life's ambition of becoming a missionary with Alan's help. Alan gave his time and brains unstintingly to his friends, paid for the schooling of a boy whom he more or less adopted, spent hours choosing suitable presents for his relations and friends, without regard to expense, and was incredibly patient with and endearing to small children, with whom he would have interesting conversations about the nature of God and other daunting subjects. All this is true to life and every word that my mother has written about it is true.

To my mind, it would have been fairer to Alan and made him more understandable to those who did not know him if my mother had mentioned some matters which contrasted sharply with his many virtues. My mother implies that his many eccentricities, divagations from normal behaviour and the rest were some kind of emanation of his genius. I do not think so at all. In my view, these things were the result of his insecurity as a child, not only in those early days at the Wards, but later on as his mother nagged and badgered him. This, however, is all theory which I am content to leave to the psychologists.

I am concerned only with Alan's behaviour as it affected other people and it was not, in my view, so amusing for those who were at the receiving end. I will give a few examples. Alan would descend upon any household at any moment of the day or night with or without warning and seldom with more than a few hours' notice. (When he was stationed at Teddington after the war, he discovered that the distance to Guildford corresponded roughly with the marathon distance, so the first we knew of his impending arrival was a badly made parcel containing a change of clothing. About twelve noon, he would come running up the steep hill of Jenner Road and straight up the stairs and into a bath.)

Alan hardly ever wrote a letter to his relatives. I understand that he carried on a huge correspondence with the other eleven mathematical pundits all over the world, the Japanese included, but all we rated was a postcard, usually arriving a day late, or the inevitable telegram. Alan was great on telegrams. "Arriving today" (not specifying exact date or time but well within the allotted shilling) was typical. My mother received loads of telegrams of this sort and they drove her wild, but nobody reading her book would think so.

Alan could not stand social chat or what he was pleased to call "vapid conversation." What he really liked was a thoroughly disputatious exchange of views. It was pretty tiring, really. You could take a safe bet that if you ventured on some self-evident proposition, as, for example, that the earth was round, Alan would produce a great deal of incontrovertible evidence to prove that it was almost certainly flat, ovular or much the same shape as a Siamese cat which had been boiled for fifteen minutes at a temperature of one thousand degrees Centigrade.

Alan's hatred of "vapid conversation," his fear of "unsafe" women and the value he placed on the importance of time – that is to say, his own – did not make him the most amiable or helpful of guests. I made the great mistake, once, of inviting him to a sherry party at my house. As a matter of fact, I find this sort of occasion pretty boring myself, but that is by the way. Alan arrived nearly an hour late, dressed like a tramp (hippies had not then been invented) and vanished within ten minutes without a word of apology or excuse. It was a silent and, perhaps, well-merited comment on our frivolous life. Frankly, I wished I could have done the same and I greatly envied him for having achieved a *modus vivendi* in which the feelings of others counted for so little.

The photographs in my mother's book are excellent and show him exactly as he was. He was a most attractive small boy: the photographs of him in his sailor suit (horrors!) are my favourites. Nor can I complain of the passage in my mother's book where she describes how he looked as a man, for the very good reason that her draft of this chapter was one of the few that she showed me and I thought her portrayal of him so preposterous that I rewrote it in less glowing terms and she actually substituted my version. All the same, I was not invited to paint him, as it were, warts and all; writing other people's scripts is, like politics, the art of the possible. Someone wrote a clerihew which ran, more or less, as follows:

> Turing
> Must have been alluring
> To get made a don
> So early on.

I won't vouch for the exact text but the point of it (which naturally my mother missed) was not so much that he was made a don at the incredibly early age of 23, but that he was – to some people, anyway – anything but alluring.

I am not, of course, referring to Alan's shabby clothes, although he might have been more alluring if they had occasionally been sent to the cleaners. He lived, as I suppose is the lot of most geniuses, in some strange world of his own, full of nervous tensions of which we lesser mortals know nothing. The manifestations of these were such as to cause consternation to the weaker brethren. For one thing, he had that painful stutter. (All stutters are so, from the late King George VI downwards, but Alan's was in a class by itself.) And, again, there was his high-pitched and raucous, unnerving, laugh, on which others have remarked. As Mrs Newman (Lyn Irvine) remarked in her Foreword to my mother's book:

> With ninety-nine people out of a hundred, Alan protected himself by his off-hand manners and his long silences – silences finally torn up by the shrill stammer and the crowing laugh which told upon the nerves of his friends. He had a strange way of not meeting the eye, of sidling out of the door with a brusque and off-hand word of thanks . . . He never looked right in his clothes, neither in his Burberry, well-worn, dirty, and a size too small, nor when he took pains and wore a clean white shirt or his best blue tweed suit.

Worst of all – and it seemed to get worse and worse as he became older – was the unsightly condition of his hands, with every finger picked raw in a dozen places. Some people never look at hands but I always do: usually one can tell more from hands than from faces. Alan's hands were those of a tormented man. As time went on, I felt sicker and sicker until I devised a special system to prevent me from looking at them at all.

It is true that Alan made a point of going about his affairs in his own inimitable way. For example, he always put his luggage in the guard's van on any journey which involved a change and left it there when he arrived at the station where he had to change. In other words, he deliberately lost it. When chided for this, he said it was less trouble and the luggage always turned up. (So it did in those days – it wouldn't do so now.)

Alan had very little notion of what went on in the great world beyond the fringe – if the expression will pass – of the academic elite. There was the most absurd and even farcical affair of his "engagement" – soon after he had been made a don so early on – to an earnest female mathematician who might be described as extremely "safe."

My parents were pretty well accustomed to my landing them with the young, attractive and lively young women with whom I fell in love at intervals of about six months at a time; they used to come for weekends and cheered up my father immensely. But Alan's fiancée was tough going. We parents and elder brother worked like beavers all the weekend on this unpromising female and were exhausted by the exercise (as, no doubt, was she). I have an improbable memory of Alan and his affianced dutifully holding hands in a sandpit, both of them obviously wishing that they could get on with some untried theorem – and that not of the type which would have appealed to me.

All this led to a peculiar correspondence which Alan referred to me for advice. The lady in question, it seems, was the only daughter of a country parson. If you are looking for a man who is after the money, give me one of those old-type country parsons. There was a short preface of seven pages or so to the effect that he could never have believed that his darling daughter's hand might be sought in holy matrimony, and, indeed, I thought he had some grounds for this notion. There were then delicate hints about money and finally the firm proposal of a marriage settlement. I daresay that my brother could have coped with the situation admirably on his own, but the fact is that he wanted to know just what a marriage settlement was, so the correspondence was referred to me. By that time, he was fed up with the poor girl and we heard no more of her.

If the episode of "the burglar" had not proved, ultimately, so fatal to Alan, I suppose this, too, might have been regarded as farcical. It occurred a couple of years before Alan's death and I believe that it was this that turned the scales against him. I had never had even the faintest notion that Alan was a homosexual. One did not in those days (at least in our middle class) talk or even think about homosexuals and lesbians: one had heard of them, of course. (There was a book called *Pansies* by D.H. Lawrence, displayed in Hatchards bookshop in Piccadilly when I was an articled clerk aged about 21. "Another boring gardening book," I sighed as I passed by.) I expect we were a little stupid.

One morning, there arrived a letter for me from Alan – a remarkable thing of itself, neither a postcard nor a telegram. I opened it and the first sentence read "I suppose you know I am a homosexual." I knew no such thing. I stuffed the letter in my pocket and read it in the office. There followed the story of "the burglar." He was not a burglar, not even a housebreaker. At this time, Alan was working at Manchester University. The so-called "burglar" was, in fact, a nasty young man whom Alan had picked up in Manchester (or the other way about) and had come by invitation to Alan's house. On his way out, he relieved Alan of his gold watch (left to him by his father) and a few other portable items. That was the "burglar" – so then, and for afterwards, described by Alan.

Alan foolishly but typically reported the loss to the police, who did not seem much interested in "the burglar." But they were greatly interested in the prospect of prosecuting a don, a near-Olympic runner and a Fellow of the Royal Society for homosexual practices, then proscribed by the law. Alan then consulted his University friends, who strongly advised him to defend the case, instruct leading counsel and heaven knows what else. In the meantime, would I kindly inform our mother of the situation? The short answer to that was that I would not.

So I dropped everything and went to Manchester where I consulted Mr G, the senior partner in a leading firm of Manchester solicitors. He, in turn, saw Alan's solicitor, Mr C, who persuaded Alan to plead guilty. In consequence, the case received the minimum of publicity. Alan was put on probation in return for an undertaking that he would undergo medical treatment. He remained secure in his university post and there were no headlines in the national press to alarm my mother. Meanwhile, he had grudgingly consented to visit her upon my insistence that she must be forewarned in case the journalists got hold of it. What exactly he told her or what she understood of it, I don't know: she did not seem to be greatly interested.

Alan did not seem to understand, even then, how close he had been to disaster, though did he, I wonder, have some premonition of things to come? He continued to talk about "the burglar" and wrote me an unpleasant letter suggesting that I cared nothing for his plight or that of homosexuals in general (the latter, perhaps, being not far wide of the mark) and that I was merely concerned to protect myself and my partners from adverse comment in the City from our Establishment friends. It was so far from the truth that I sent him a tart reply of which I feel ashamed. It was a disagreeable business and the only occasion I can remember that we quarrelled.

Some two years later, during the Whitsun holiday, I had taken one of my daughters to the cinema and arrived home about 10.30pm. In my absence, the Manchester police had telephoned to say that Alan had been found dead in his house. Late as it was, I telephoned the ever kindly and shrewd Mr G, who promised to meet me at the station in Manchester the next morning. He took me to the police and thence we went to the mortuary where I identified Alan's body. He had taken cyanide. By great good fortune, my mother was on holiday in Italy and did not return home until after the inquest.

Mr G advised me strongly not to instruct counsel to appear at the inquest and told me of the unhappy course which some other cases had taken before this coroner, a retired doctor who could not abide lawyers. The possibility of establishing death by accident was minimal; the best we could hope for was the considerate verdict of "balance of mind disturbed." He was right and I accepted his advice. At the inquest itself, this soon became apparent: there were present some eight or nine reporters, some from the national press, with pencils poised and waiting for the homosexual revelations. They were disappointed. I gave evidence briefly. The coroner asked me a few perfunctory questions. The verdict was as anticipated.

When my mother returned, she was highly indignant and made no secret of her belief that I had grossly mishandled the case. She evolved various theories of her own to establish to her own satisfaction that it was really an accident. But I had worked on the very same theories myself in Manchester for nearly three days and there was one fatal flaw in them. This was the half-eaten apple beside Alan's bed where his body was found. The apple was to disguise the bitter taste of the cyanide and thus ensure that the poison would do its work.

In those unhappy days in Manchester, I visited Alan's psychiatrist who told me a great deal about Alan that I did not know before – among other things, that he loathed his mother. I refused to believe it. He then handed me two exercise books in which Alan had entered such matters as psychiatrists require of their patients, including their dreams. "You had better take them away and read them," he said, adding that there was a third book, probably in Alan's house.

I viewed the two books in my hotel with horror, but I was still bent on proving the accident theory and decided I had better read them. I wish I had not. Alan had been a practising homosexual since the age of puberty. His comments on his mother were scarifying. To my great relief, I was mentioned only once or twice and not in opprobrious terms. I returned the books to the psychiatrist the following day. There remained the problem of the third book, for it was essential that it should be found so that it would not fall into my mother's hands. Eventually it was found and returned to the psychiatrist. Two days later, my mother arrived in Manchester and ransacked the house for clues bearing on her preconceived theories. I need hardly add that she remained unaware of the books and of Alan's feelings about her until the day of her death.

I once asked my father what he hated most in the world. He replied, without a moment's hesitation, "Humbug." My mother has dispensed a liberal dose of it to the general public. Looking back on my father and brother as I remember them, I think I owe it to their memories to put the record straight. Inevitably, it has fallen to my lot to present some less appealing features of Alan's character and habits. He was a complex man and much loved by many. Had he been better understood when he was young – and if I, among others, had treated him with more consideration – he might be alive today. Why he took his own life remains, and is likely to remain, an unsolved mystery. I take the view, myself, that the immense pressures on him at Bletchley for so many years had taken their toll and that, but for these, he might have weathered these later storms.

If I have dwelled too much on Alan's idiosyncracies, it is because these were, to me, the essential Alan, but to many who saw a different side of him, he is remembered more for his modesty, his generosity and his kindness. I cannot do better than to quote from Mr O'Hanlon's notice in the school magazine, *The Shirburnian*, after Alan's sad death in June 1954:

For those who knew him here, the memory is of an even-tempered, loveable character with an impish sense of humour and a modesty proof against all achievement. You would not take him for a wrangler, the youngest Fellow of King's and almost the youngest FRS of his time, or a marathon runner, or that behind a negligé appearance, he was intensely practical . . . In all his preoccupation with logic, mathematics and science, he never lost the common touch; in a short life he accomplished much, and to the roll of great names in the history of his particular studies added his own.[5]

---

[5] Quoted by kind permission of Mr. Michael Hanson.

BIBLIOGRAPHY

~

Bentham, G. *Handbook of the British Flora.* Revised by Sir J.D. Hooker and A.B. Rendle. L. Reeve & Co., Ashford, Kent. 1947. [Page 102.]

Brewster, E.T. *Natural Wonders Every Child Should Know.* Grosset & Dunlap, New York. 1912. [Page 20.]

Butler, Samuel. *Erewhon.* Everyman. 1919. [Pages 108, 132.]

Clapham, A.R., Tutin, T.G., Warburg, E.F. *Flora of the British Isles.* Cambridge University Press. 1952. [Page 102.]

Eddington, Sir A. *The Nature of the Physical World.* Cambridge University Press and Macmillan Co., New York. 1928. [Page 83.]

German Mathematical Encyclopaedia. (Encyklopadie der Mathematischen Wissenschaften.) Teubner, Leipzig. 1898/1904–1935. Revised edn. 1939. [Page 48.]

Hobson, E.W. *Spherical and Ellipsoidal Harmonics.* Cambridge University Press. 1931. [Page 138.]

Neumann, J. VON. *Mathematische Grundlagen der Quantenmechanik.* Springer, Berlin. 1932. [Page 38.]

Newman, J.R., Ed. *The World of Mathematics.* Vols. II and IV. Simon & Schuster, New York. 1956. [Pages 93, 94, 125–127.]

Pringle, J.W.S. The Origin of Life. *Symposia of the Society for Experimental Biology.* Number VII. *Evolution.* 1953. [Pages 133–134.]

Smith, R.C. *Eyes and No Eyes.* Cassel, London. 1924. [Page 14.]

Stoney, F.S. *Some Old Annals of the Stoney Family.* [Page 5.]

Thompson, Sir D'arcy W. *Growth and Form.* New edn., 2 vols. Cambridge University Press. 1952. [Page 135.]

Wiener, Norbert. *Cybernetics.* John Wiley & Son. Inc., New York. 1948. [Page 93.]

The Human Use of Human Beings. Houghton Mifflin, Boston, Mass. 1950. Also Eyre & Spottiswoode, London. 1950. [Page 93.]

Turing, A.M. 1935*a*. On the Gaussian Error Function (King's College Fellowship Dissertation).   [Pages 44, 45.]

1935*b*. Equivalence of left and right almost periodicity, *J. Lond. Math. Soc.*, **10**, 284.   [Pages 44, 120.]

1937*a*. On computable numbers, with an application to the Entscheidungsproblem, *Proc. Lond. Math. Soc.* (2), **42**, 230. [Pages, 46, 47, 48, 52–54, 77, 98.]

1937*b*. Computability and λ-definability, *J. Symbolic Logic*, **2**, 153. [Page 48.]

1937*c*. The *p*-function in λ–*K*-conversion, *J. Symbolic Logic*, **2**, 164. [Page 48.]

1937*d*. Correction to 1937*a*, *Proc. Lond. Math. Soc.* (2), **43**, 544. [Page 46.]

1938*a*. Finite approximations to Lie groups, *Ann. Math., Princeton*, **39**, 105.   [Pages 52, 120.]

1938*b*. The extensions of a group, *Compos. Math.*, **5**, 357.   [Page 52.]

1939. Systems of logic based on ordinals, *Proc. Lond. Math. Soc.* (2), **45**, 161.   [Pages 13, 54.]

1942*a*. (With M.H.A. NEWMAN.) A formal theorem in Church's theory of types, *J. Symbolic Logic*, **7**, 28.   [Page 55.]

1942*b*. The use of dots as brackets in Church's system, *J. Symbolic Logic*, **7**, 146.   [Page 55.]

1943. A method for the calculation of the zeta-function, *Proc. Lond. Math. Soc.* (2), **48**, 180. (Received four years earlier (7th March, 1939).)   [Page 55.]

1948*a*. Rounding-off errors in matrix processes, *Quart. J. Mech. App. Math.*, **1**, 287.   [Page 85.]

1948*b*. Practical forms of type-theory, *J. Symbolic Logic*, **13**, 80. [Page 85.]

1950*a*. The word problem in semi-groups with cancellation, *Ann. Math., Princeton*, **52**, 491.   [Pages 54, 98.]

1950*b*. Computing machinery and intelligence, *Mind*, **59**, 433. (Reprinted with the title "Can a machine think?" *The World of Mathematics*. Vol. IV. Ed. J.R. Newman. Simon & Schuster, New York. 1956.)   [Pages 94, 95.]

1950c. *Programmers' Handbook for the Manchester electronic computer.* [Page 88.]

1952. The chemical basis of morphogenesis, *Phil. Trans. B*, **237**, 37. [Page 102.]

1953a. Some calculations of the Riemann zeta-function, *Proc. Lond. Math. Soc.* (3), **3**, 99.   [Page 98.]

1953b. Digital computers applied to games: chess, pp. 288–295 of *Faster than Thought*, ed. B.V. Bowden. Pitman, London.   [Page 97.]

1954. Solvable and unsolvable problems, *Sci. News*, **31**, 7. [Pages 97, 98.]

On a theorem of Littlewood. In *Collected Works*. A.M. Turing. North-Holland Publishing Co., Amsterdam. 1959.   [Page 48.]

Intelligent machinery. (Report for N.P.L. 1947–48). In *Collected Works*. A.M. Turing. North-Holland Publishing Co., Amsterdam. 1959. [Page 87.]

The chemical theory of morphogenesis, Part II, and the Morphogen theory of phyllotaxis. (Prepared for publication by N.E. Hoskin and B. Richards.) In *Collected Works*. A.M. Turing. North-Holland Publishing Co., Amsterdam. 1959.   [Page 135.]

Intelligent machinery. A heretical theory. *This volume.* [Pages 127–132.]